Advanced Research in Plant Genetics

Advanced Research in Plant Genetics

Edited by Herbert McCoy

SYRAWOOD
PUBLISHING HOUSE

New York

Published by Syrawood Publishing House,
750 Third Avenue, 9th Floor,
New York, NY 10017, USA
www.syrawoodpublishinghouse.com

Advanced Research in Plant Genetics
Edited by Herbert McCoy

Cataloging-in-Publication Data

Advanced research in plant genetics / edited by Herbert McCoy.
    p. cm.
Includes bibliographical references and index.
ISBN 978-1-68286-855-3
1. Plant genetics. 2. Plant genetics--Research. 3. Plant genetics--Technique. I. McCoy, Herbert.
QK981 .A28 2020
581.35--dc23

TABLE OF CONTENTS

PREFACE

Plant genetics is associated with the study of genes, heredity and genetic variations in plants. It is considered a field of botany and biology, but frequently intersects with other life sciences as well. There are major economic impacts of the study of plant genetics such as genetic modification of staple crops to increase their yield, confer pest and disease resistance, or provide resistance to herbicides. An understanding of the mode of reproduction is vital for facilitating the manipulation and breeding of improved plant varieties. Research in modern plant genetics and the sequencing of many plant genomes are the results of genetic modification. *Agrobacterium* method and gene gun are the two predominant procedures for transforming genes in organisms. The various studies that are constantly contributing towards advancing technologies and the evolution of plant genetics are examined in detail in this book. Students, researchers, experts and all associated with this discipline will benefit alike from this book.

The researches compiled throughout the book are authentic and of high quality, combining several disciplines and from very diverse regions from around the world. Drawing on the contributions of many researchers from diverse countries, the book's objective is to provide the readers with the latest achievements in the area of research. This book will surely be a source of knowledge to all interested and researching the field.

In the end, I would like to express my deep sense of gratitude to all the authors for meeting the set deadlines in completing and submitting their research chapters. I would also like to thank the publisher for the support offered to us throughout the course of the book. Finally, I extend my sincere thanks to my family for being a constant source of inspiration and encouragement.

Editor

GENETIC DIVERSITY AND PHENETIC RELATIONSHIPS OF FIVE *TRIFOLIUM* L. SPECIES (FABACEAE) BY INTER SIMPLE SEQUENCE REPEATS MARKERS

YOURANG HWANG AND MAN KYU HUH[1]

Department of Molecular Biology, Dong-eui University, 995 Eomgwangno, Busanjin-gu, Busan 614-714, Korea

Keywords: Genetic variation; Inter simple sequence repeats; *Trifolium.*

Abstract

Five species of *Trifolium* L. (*T. repens* L., *T. pretense* L., *T. hybridum* L., *T. campestre* Schreb., and *T. dubium* Sibth.) were analyzed used to evaluate the genetic diversity and their phenetic relationships using inter-simple sequence repeats (ISSR) markers. Overall, *T. pratense* exhibited higher variation than other species. 114 amplicons were produced by ISSR markers, of which 77 (67.5%) bands were polymorphic. *T. dubium* showed the low genetic variation. Total genetic diversity values (H_T) varied between 0.333 and 0.487, for an average over all polymorphic loci of 0.282. On a per-locus basis, the proportion of total genetic variation due to differences among species (G_{ST}) was 0.380. This indicated that about 38.0% of the total variation was among species. The estimate of gene flow, based on G_{ST}, was very low among species of genus *Trifolium* (N_m = 0.816). An assessment of the proportion of diversity present within species, H_{POP}/H_{SP}, indicated that about 95.8% the total genetic diversity was within species. *T. pratense* and *T. hybridum* were grouped together and this clade was sister with *T. repens*. Two remainder species with yellow flowers were grouped together. Information on genetic diversity for *Trifolium* is valued for the management of germplasm and for evolving conservation strategies.

Introduction

Trifolium L., the clover genus, is one of the largest genera in Fabaceae family. This genus consists about 250-300 species with a wide distribution and adaption to different agro-ecological regions (Gillet *et al.*, 2001; Ellison *et al.*, 2006). *Trifolium repens*, also known as White or Dutch clover, originated in the Mediterranean region and quickly spread throughout Europe (Baker and Williams, 1987; Lane *et al.*, 1997). White clover is adapted to a wide climate range from the Arctic to the subtropics and has a wide altitudinal ranges. It is found up as 6000 m in the Himalaya Range (Baker and Williams, 1987). It has also become naturalized in China, Mongolia, Korea, and Japan.

The genus *Trifolium* includes more than 20 clover cultivated species as forages (Hirano, 2005). White clover (*T. repens*) in Korea has been introduced from Europe about two hundred years ago. Most species belonging to genus *Trifolium* can tolerate wide variations in temperature, sunlight, and pH of soil. With the recent development of organic farming, legumes have been considered candidates of fertilizer (Paplauskiene and Dabkeviciene, 2012). Many species of *Trifolium* are known to have been cultivated on a commercial scale including white and red clover (*T. repens* and *T. pratense*), the two most economically important pasture legumes in the UK (Taylor and Quesenberry, 1996). However, *Trifolium* is one of major weeds for lawns, farming fields, and golf courses in Korea. Especially many plants of *Trifolium* are also considered to cause

---

[1] Corresponding author. E-mail: mkhuh@deu.ac.kr

damage to the environment and have gradually the superior competitive ability on golf courses to create fairways and teeing areas.

Alsike clover (*T. hybridum*), field clover (*T. campestre*), suckling clover   (*T. dubium*), are European grassland legumes that have spread to many parts of the world. Recently they have been also introduced to Korea. These non-native clovers can rapidly invade and dominate vegetated and bare areas in Korea (Huh *et al.*, 2005).

Many molecular marker techniques have been developed and they have been extensively used in plant systematic studies, measurement of variation to establish evolutionary relationships within or among species, and population genetic research (Hu and Vick, 2003; Gupta and Rustgi, 2004; Rizza *et al.*, 2007). Inter simple sequence repeats (ISSR) markers have the advantage over randomly amplified polymorphic DNA (RAPD) in that the primers are longer, allowing for more stringent annealing temperatures (Wolfe and Liston, 1998). These higher temperatures apparently provide a higher reproducibility of bands than in RAPD (Nagaoka and Ogihara, 1997). Tsumura *et al.* (1996) found that most of their ISSR bands (96%) segregated according to Mendelian expectations.

The aim of this study was the estimation of population structure, genetic diversity, and genetic relationships of five clover species in Korea.

Materials and Methods

Plant materials

Five clover species, *T. repens* L., *T. pratense* L., *T. hybridum* L., *T. campestre* Schreb., and *T. dubium* Sibth. were used for ISSR analysis (Table 1). Thirty plants were collected for each species. Within populations, plants are genetically subdivided by micro-environmental heterogeneity (Sackville and Chorlton, 1995). Clover has a creeping growth habit and spreads with stolons or runners above the soil with adventitious roots forming at each node. The geographic distance between the selected individuals was about 1.0 m to avoid inclusion of individuals emanating from the same rhizome. *Medicago sativa* L. was used as an outgroup species in this study.

DNA extraction

Total genomic DNA was extracted from a fresh young leaves using the plant DNA Zol Kit (Life Technologies Inc., Grand Island, New York, U.S.A.) according to the manufacturer's protocol. Briefly, Approximately 1.2 g fresh leaves per individual was ground to fine powder in liquid nitrogen with a mortar and pestle. The pulverized material was transferred to a micro-tube and Plant DNA Zol solution was added. The sample was shaken gently at room temperature for 10 min. After adding 24:1 chloroform/isoamyl alcohol, the sample was centrifuged at 12,000 g. The DNA precipitate was recovered with 70% ethanol, dried, and dissolved in TE buffer. The extracted DNA concentrations were calculated with a fluorometer (DyNA Quant 200, Hoefler, Amersham Biosciences, USA) using bisbenzimide (Amersham Biosciences, USA) as the fluorescent dye.

ISSR analysis

The ISSR amplification assay developed by Zietkiewicz *et al.* (1994) using primers listed in Table 1. PCR was performed within a total volume of 25 □ using a PTC-100 DNA Engine Dyad Peltier thermal cycler (MJ Research, Watertown, MA, USA). Each PCR mixture contained PCR buffer (Promega; 20 mM This-HCl, 50 mM KCl), 1.5 mM $MgCl_2$, 0.24 mM of each dNTP, 12.5 pmol of each primer, 0.25 units of BIOTAQ DNA polymerase (Bioline), and 25 ng of genomic

DNA. An initial denaturation step of 5 min at 94°C was followed by 30 cycles of amplification (1 min sec at 94°C, 1 min at 50°C, 1.5 min at 72°C) and a final elongation step of 10 min at 72°C.

The amplification products were separated by electrophoresis on 2.0% agarose gels in Tris-Borate buffer, and stained with ethidium bromide. A 100 bp ladder DNA marker (Pharmacia) was used in the end of for the estimation of fragment size.

Statistical analyses

PCR-amplified ISSR fragments detected on gels were scored absent (0) or present (1). Only unambiguously reproducible bands were scored and used in the analyses. The following genetic parameters were calculated using a POPGENE computer program (ver. 1.31) developed by Yeh *et al.* (1999): the percentage of polymorphic loci (P_p), mean numbers of alleles per locus (A), effective number of alleles per locus (A_e), and gene diversity (H) (Nei, 1973) and Shannon's index (I) of phenotypic diversity. Shannon–Weaver index of diversity (Shannon and Weaver, 1963): the formula for calculating the Shannon diversity index (H') is:

$$H' = - \Sigma \, pi \ln pi$$

pi is the proportion of important value of the ith species ($pi = ni \,/\, N$, ni is the important value index of ith species and N is the important value index of all the species).

Polymorphism information content (PIC) value was calculated using the formula PIC, PIC = 1 - p^2- q^2, where p = band frequency and q = no-band frequency (Rizza *et al.*, 2007).

Nei's gene diversity formulae (H_T, H_S, and G_{ST}) were used to evaluate genetic diversity within and among cultivars (Nei, 1973). H_T is the expected heterozygosity of an individual in an equivalent random mating total interspecies. H_S is the expected heterozygosity of an individual in an equivalent random mating total intraspecies. The G_{ST} coefficient corresponds to the relative amount of differentiation among cultivars. Furthermore, gene flow (*Nm*) between the pairs of species was calculated from G_{ST} values by $Nm = 0.5(1/G_{ST} - 1)$ (McDermott and McDonald, 1993).

Shannon's index of genotypic diversity (H_O) for ISSR was estimated to quantity the degree of within species diversity following the formula (Bowman *et al.*, 1971): $H_O = -\Sigma pi \log pi$, where *pi* is the frequency of a particular phenotype *i*.

A phenetic relationship was constructed by the neighbor-joining (NJ) method in PHYLIP version 3.57 using MEGA5 program (Tamura *et al.*, 2011). Parsimony analyses were conducted using PAUP* *4.0b3a* (Swofford, 1999). Confidence values for individual branches were determined by a bootstrap analysis with 100 repeated sampling of the data.

Results and Discussion

From the 20 decamer primers used for a primary ISSR analysis, thirteen primers produced good amplification products both in quality and variability (Table 1). The remaining primers either did not amplify or showed unclear amplification across all genotypes. 114 amplicons were produced by ISSR marker, of which 77 (67.5%) bands were polymorphic. Polymorphism information content (PIC) for ISSR primers ranged from 0.244 to 0.498 with an average of 0.287 per primer.

In a simple measure of inter-cultivars variability i.e. the percentage of polymorphic bands, *T. pratense* exhibited the highest variation (49.1%) among clovers and *T. dubium* the lowest (36.0%) (Table 2). The average number of alleles per locus (A) was 1.423 across species, varying from 1.360 to 1.491. The effective numbers of alleles per locus (A_E) was 1.311 across species, varying from 1.251 to 1.374. The mean genetic diversity within species was 0.175. Shannon's index of phenotypic diversity (*I*) of *T. pratense* (0.302) was highest of all taxa and *T. hybridum* was the second (0.286). Overall, *T. pratense* exhibited higher variation than other species. Two species (*T. campestre* and *T. dubium*) with yellow flowers were shown the low genetic variation.

The first fragment (ISSR-06-01) of primer ISSR-01 was specific band for *T. repens* which did not show at other species. The ISSR-01-04 fragment of primer ISSR-01 was also specific band for *T. pratense*. These specific fragments seemed to be useful markers to discriminate among species.

Table 1. List of decamer oligonucleotide utilized as primers, their sequences, and associated polymorphic fragments.

No. of Primer	Sequence(5' to 3')	No. of fragments detected	Percentage of polymorphism bands	PIC
ISSR-01	$(AG)_8G$	11	9	0.489
ISSR-02	$(CA)_8RG$	8	8	0.498
ISSR-03	$(GA)_8GT$	10	7	0.458
ISSR-04	$(GA)_8CG$	7	4	0.452
ISSR-05	$(GA)_8GT$	10	7	0.328
ISSR-06	$(GA)_8CG$	11	7	0.418
ISSR-07	$(GA)_8TC$	9	6	0.452
ISSR-08	$(GA)_8TC$	7	5	0.408
ISSR-09	$GCGA(AC)_8$	9	7	0.285
ISSR-10	$GCGA(CA)_8$	7	4	0.328
ISSR-11	$CCGG(AC)_8$	10	5	0.275
ISSR-12	AGAGTTGGTAGCTCTTG ATC	8	4	0.244
ISSR-13	$(AC)_8T$	7	4	0.310
Total	-	114	77	0.287

Table 2. Measurements of genetic variation for five clover species used in this study. The number of polymorphic loci (Np), percentage of polymorphism (Pp), mean number of alleles per locus (A), effective number of alleles per locus (A_E), gene diversity (H), and Shannon's information index (I).

Species	N_P	P_P	A	Ae	H	I
Trifolium repens	48	42.1	1.421	1.309	0.173	0.252
Trifolium hybridum	53	46.5	1.465	1.360	0.198	0.286
Trifolium pratense	56	49.1	1.491	1.374	0.209	0.302
Trifolium campestre	43	37.7	1.377	1.251	0.147	0.217
Trifolium dubium	41	36.0	1.360	1.259	0.147	0.215
Mean	48.2	42.3	1.423	0.311	0.175	0.254

Total genetic diversity values (H_T) for polymorphic loci varied between 0.333 (ISSR-04) and 0.487 (ISSR-12) (Table 3). An average (H_T) over all 114 loci for five species with 13 ISSR primers was 0.282. In interlocus variation in the within-species, mean genetic diversity (H_S) was low (0.175). On a per-locus basis, the proportion of total genetic variation due to differences among species (G_{ST}) ranged from 0.216 for ISSR-08 to 0.547 for ISSR-02, with a mean of 0.380. This indicated that about 38.0% of the total variation was among species. Thus, about genetic variation (62.0%) resided within species. The estimate of gene flow, based on G_{ST}, was very low among species (N_m = 0.816). Values of genetic distance (D) were ≤ 0.233 (Table 4). Genetic identity values among pairs of species ranged from 0.508 to 0.956.

Table 3. Estimates of genetic diversity of five selected clover species in Korea. Total genetic diversity (H_T), genetic diversity within populations (H_S), the proportion of total genetic diversity partitioned among populations (G_{ST}), and gene flow (Nm).

Primer	H_T	H_S	G_{ST}	Nm
ISSR-01	0.405	0.231	0.422	1.821
ISSR-02	0.428	0.186	0.547	0.778
ISSR-03	0.380	0.217	0.420	1.097
ISSR-04	0.333	0.199	0.304	5.631
ISSR-05	0.409	0.197	0.527	3.061
ISSR-06	0.395	0.264	0.358	7.133
ISSR-07	0.432	0.294	0.324	4.266
ISSR-08	0.474	0.368	0.216	5.405
ISSR-09	0.386	0.265	0.331	2.279
ISSR-10	0.453	0.338	0.242	2.880
ISSR-11	0.474	0.317	0.328	2.791
ISSR-12	0.487	0.367	0.248	2.917
ISSR-13	0.427	0.247	0.413	0.871
Total mean	0.282	0.175	0.380	0.816

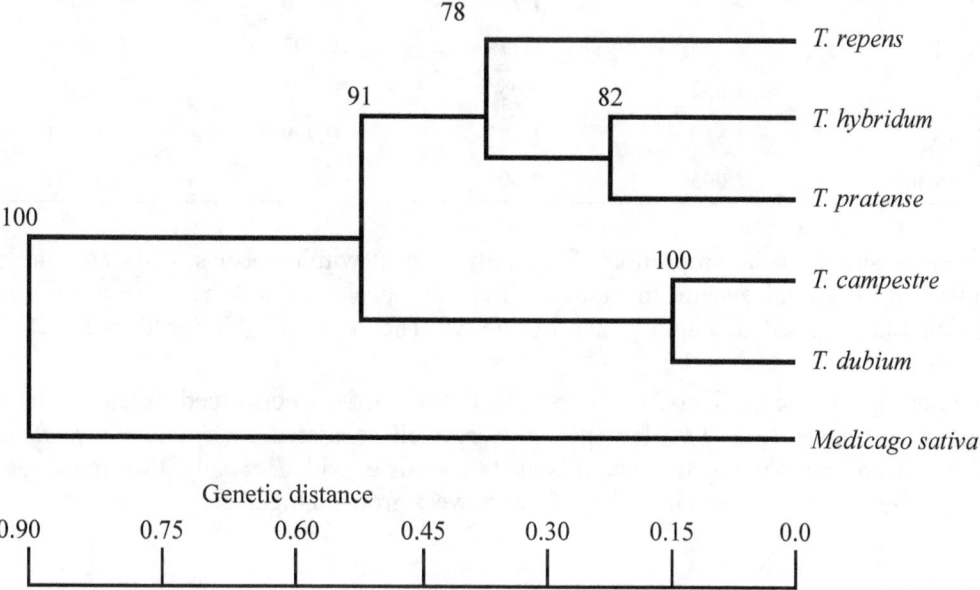

Fig. 1. A phentic tree for five selected species with one outgroup based on ISSR analysis. Numbers above branches are jackknife values derived from heuriatic-based searches on sequences data.

Table 4. Genetic identity (upper diagonal) among five selected clover species and genetic distances (low diagonal) based on ISSR analysis.

Species	T. repens	T. hybridum	T. pratense	T. campestre	T. dubium
T. repens	-	0.815	0.849	0.815	0.807
T. hybridum	0.205	-	0.866	0.508	0.792
T. pratense	0.164	0.144	-	0.833	0.836
T. campestre	0.205	0.217	0.183	-	0.956
T. dubium	0.214	0.233	0.179	0.045	-

Table 5. Partitioning of the genetic diversity into within and among genus _Trifolium_ in Korea.

Primer	H_{VAR}	H_{SP}	H_{VAR} / H_{SP}	$(H_{SP} - H_{VAR})/ H_{SP}$
ISSR-01	2.110	2.306	0.915	0.085
ISSR-02	1.874	1.901	0.986	0.014
ISSR-03	2.057	2.207	0.932	0.068
ISSR-04	1.660	1.806	0.919	0.081
ISSR-05	2.209	2.277	0.970	0.030
ISSR-06	2.187	2.317	0.944	0.056
ISSR-07	2.043	2.138	0.955	0.045
ISSR-08	1.886	1.914	0.986	0.014
ISSR-09	2.109	2.182	0.967	0.033
ISSR-10	1.771	1.912	0.926	0.074
ISSR-11	2.263	2.279	0.993	0.007
ISSR-12	2.052	2.064	0.994	0.006
ISSR-13	1.848	1.907	0.969	0.031
Total mean	2.005	2.093	0.958	0.042

An assessment of the proportion of diversity present within species, H_{VAR}/H_{SP}, indicated that about 95.8% the total genetic diversity was within species. Thus, the other portion of genetic variation (4.2%) resided within genus (Table 5). The result was lower than that (G_{ST}) of F-statistics.

Clustering of five cultivars, using the NJ algorithm, was performed based on the matrix of calculated distances (Fig. 1). Five species were well separated each other. _T. pratense_ and _T. hybridum_ were grouped together and this clade was sister with _T. repens_. Two remainder species, _T. campestre_ and _T. dubium_ with yellow flowers were grouped together.

References
Baker, M.J. and Williams, W.M. 1987. White Clover. CAB International, Wallingford, UK, pp. 299-322.

Bowman, K.D., Hutcheson, K., Odum, E.P. and Shenton, L.R. 1971. Comments on the distribution of indices of diversity. Stat. Ecol. **3**: 315-359.

Ellison, N.E., Liston, A., Steiner, J.J., Williams, W.M. and Taylor, N.L. 2006. Molecular phylogenetics of the clover genus (*Trifolium-Leguminosae*). Mol. Phylogenet. Evol. **39**: 688-705.

Gillet, J.M., Collins, M. and Taylor, N.J. 2001. The World of Clovers. Iowa State University Press, Ames., pp. 457.

Gupta, P.K. and Rustgi, S. 2004. Molecular markers from the transcribed/expressed region of the genome in higher plants. Funct. Integr. Genomics **4**: 139-162.

Hirano, R. 2005. Ecogeographic and genetic survey of white clover (*Trifolium repens* L.) on St Kilda. Thesis of Master, University of Birmingham, UK, pp. 1-85.

Hu, J. and Vick, B.A. 2003. Target region amplification polymorphism: a novel marker technique for plant genotyping. Plant Mol. Biol. Report **21**: 289-294.

Huh, M.K., Chung, K.T., and Jeong, Y.K. 2005. Genetic variation of alien invasive red clover (*Trifolium pratense*) in Korea. J. Life Sci. **15**: 273-278.

Lane, L.A., Ayes, J.F. and Lovett, J.V. 1997. A review of the introduction and use of white clover (*Trifolium repens* L.) in Australia – significance for breeding objectives. Australian Journal of Experimental Agriculture **37**: 831-839.

Nagaoka, T. and Ogihara, Y. 1997. Applicability of inter-simple sequence repeat polymorphisms in wheat for use as DNA markers in comparison to RFLP and RAPD markers. Theor. Appl. Genet. **94**: 597-602.

McDermott, J.M. and McDonald, B.A. 1993. Gene flow in plant pathosystems. Ann. Rev. Phytopathy. **31**: 353-373.

Nei, M. 1973. Analysis of gene diversity in subdivided populations. Proc. Natl. Acad. Sci. USA **701**: 3321-3323.

Paplauskienė, V. and Dabkevičienė, G. 2012. A study of genetic diversity in *Trifolium hybridum* varieties using morphological characters and ISSR markers. Žemdirbystė-Agriculture, **99**: 313-318.

Rizza, M.D., Real, D., Reyno, R., Porro, V., Burgueno, J., Errico, E. and Quesenberry, K.H. 2007. Genetic diversity and DNA content of three South American and three Eurasiatic *Trifolium* species. Genetics and Molecular Biology **30**: 1118-1124.

Sackville, H. and Chorlton, N.R. 1995. Collecting Plant Genetic Diversity. CAB International, Wallingford, Oxford, pp. 467-483.

Shannon, C. E. and Weaver, W. 1963. The Measurement Theory of Communication. Univ. of Illinois Press, Urbana, 1-132.

Swofford, D. L. 1999. PAUP*. Phylogenetic analysis using parsimony (*and other methods). Ver. 4.0b3a. Sinauer Associates,Sunderland, Massachusetts.

Tamura, K., Peterson, D., Peterson, N., Stecher, G., Nei, M. and Kumar, S. 2011. MEGA5: Molecular evolutionary genetics analysis using maximum likelihood, evolutionary distance, and maximum parsimony methods. Mol. Biol. Evol. **28**: 2731-2739.

Taylor, N.L. and Quesenberry, K.H. 1996. Red Clover Science, Kluwer, Boston, MA. pp. 11-24.

Tsumura, Y., Ohba, K. and Strauss, S.H. 1996. Diversity and inheritance of inter-simple sequence repeat polymorphisms in Douglas-fir (*Pseudotsuga menziesii*) and sugi (*Cryptomeria japonica*). Theor. Appl. Genet. **93**: 40-45.

Wolfe, A.D. and Liston, A. 1998. Molecular Systematics of Plants: DNA Sequencing, Kluwer, New York, USA, pp. 43-86.

Yeh, F.C., Yang, R.C. and Boyle, T. 1999. POPGENE Version 1.31, Microsoft Windows-based Freeware for Population Genetic Analysis. University of Alberta, Alberta, pp. 1-228.

Zietkiewicz, E., Rafalski, A. and Labuda, D. 1994. Genome fingerprinting by simple sequence repeat (SSR)-anchored polymerase chain reaction amplification. Genomics **20**: 176-183.

2

SEED GERMINATION BEHAVIOUR OF SIX MEDICINAL PLANTS FROM BANGLADESH

ALEYA FERDOUSI, MD. OLIUR RAHMAN[1] AND MD. ABUL HASSAN

Department of Botany, University of Dhaka, Dhaka-1000, Bangladesh

Keywords: Medicinal plants; Seed germination; Bangladesh.

Abstract

This paper focuses on seed germination of six indigenous medicinal plants of Bangladesh, namely *Adenanthera pavonina* L., *Helicteres isora* L., *Murraya paniculata* (L.) Jack, *Psoralea corylifolia* L., *Uraria lagopodioides* (L.) Desv. and *U. picta* (Jacq.) Desv. *ex* DC. The minimum days taken to germinate seeds in *Adenanthera pavonina* L., *Murraya paniculata* (L.) Jack, *Psoralea corylifolia* L., *Uraria lagopodioides* (L.) Desv. and *U. picta* (Jacq.) Desv. *ex* DC. are 12, 36, 10, 39 and 14, respectively. Seeds were not germinated in *Helicteres isora* L. indicating that seeds are not suitable for propagation, however, propagation through stem cutting in this species revealed that plants flowers and set fruits in the same year and take only six to seven months. Epigeal type of seed germination was observed in all cases.

Introduction

Medicinal plants play an important role in human life since they are employed as raw materials for the extraction of active constitution in pure form, as precursor for synthetic vitamins and steroids, and as preparations for herbal and indigenous medicines (de Padua *et al.*, 1999). Yusuf *et al.* (2009) documented 747 species of medicinal plants occurring in Bangladesh. *Adenanthera pavonina* L., *Helicteres isora* L., *Murraya paniculata* (L.) Jack, *Psoralea corylifolia* L., *Uraria lagopodioides* (L.) Desv. and *U. picta* (Jacq.) Desv. *ex* DC. are six important medicinal plants commonly found in the country and used in traditional medicine. Seeds of *Adenanthera pavonina* L. (Fabaceae) are used in the treatment of boils, inflammation, cholera and paralysis (Ghani, 2003). Leaf paste of *Helicteres isora* L. (Sterculiaceae) is used in the treatment of eczema, while stem bark and roots are considered to be demulcent, expectorant, astringent and anti-galactagogue, and are employed for treating dysentery, diarrhoea and biliousness (Ghani, 2003). Leaves of *Murraya paniculata* (L.) Jack (Rutaceae) are astringent, and used in diarrhoea and dysentery; a decoction of leaves is taken in dropsy and powdered leaf is applied to fresh cuts (Yusuf *et al.*, 2009). *Psoralea corylifolia* L. (Fabaceae) is claimed to be useful in skin disorders, eczema and hair loss; fruits are laxative, aphrodisiac and are used for the treatment of leucoderma and leprosy; while seeds are used as laxative, diaphoretic, stomachic and anthelmintic (Ghani, 2003). *Uraria lagopodioides* (L.) Desv. is used in remittent fever, asthma, dysentery and for treatment of inflammation in chest. Decoction of leaves is used in diarrhoea (Yusuf *et al.*, 2009). *U. picta* (Jacq.) Desv. *ex* DC. a source of antiseptic and leaves are used in gonorrhoea; roots are aphrodisiac and decoction of roots is used in fever and cough (Yusuf *et al.*, 2009).

The germination response pattern of seeds is an important phenomenon in plant life history strategy (Mayer and Poljakoff-Mayber, 1989). In the recent past studies on seed germination and reproductive biology on different groups of plants have received considerable attention (Chauhan and Johnson, 2008; Liebst and Schneller, 2008; Vandelook and van Assche, 2009; Clements *et al.*, 2010; Han and Long, 2010; Kameneva and Koksheeva, 2013), however, very little is known on

---

[1]Corresponding author. E-mail: prof.oliurrahman@gmail.com

the seed germination pattern of medicinal plants (Hassan and Fardous, 2003; Liza *et al.*, 2010; Rahman *et al.*, 2012). Since medicinal plants are employed for primary healthcare system, emphasis to be given on seed germination patterns of medicinal plants, as in many cases they need to bring under cultivation. However, no earlier study has surveyed germination patterns in the medicinal plants employed in the present study. Therefore, the objective of the present work is to explore seed germination pattern and dormancy of seeds in *Adenanthera pavonina* L., *Helicteres isora* L., *Murraya paniculata* (L.) Jack, *Psoralea corylifolia* L., *Uraria lagopodioides* (L.) Desv. *ex* DC. which might help in bringing the plants under cultivation.

Materials and Methods

Six medicinally important plants selected for this study are *Adenanthera pavonina* L., *Helicteres isora* L., *Murraya paniculata* (L.) Jack, *Psoralea corylifolia* L., *Uraria lagopodioides* (L.) Desv. and *U. picta* (Jacq.) Desv. *ex* DC. Plants materials were collected from different areas of the country and planted in the Botanical Garden of Dhaka University for closer observation and critical study. The voucher specimens are deposited in Dhaka University Salar Khan Herbarium (DUSH).

Seeds of six species were collected from mature fruits and preserved under laboratory condition. Rahman *et al.* (2012) was followed for seed germination experiment. For sowing of the seeds earthen pots of 10 inch in diameter filled up with a mixture of soil and compost (2:1). In order to prevent fungal infection and microbial contamination seeds were treated with fungicides prior to sowing. Ten mature seeds for each taxon were sown in earthen pots at different time intervals to record dormancy and viability, suitable time for germination, percentage and type of germination. Propagation through stem cutting was performed in *Helicteres isora* as seeds were not germinated in this species.

Result and Discussion

Seed germination study on six species revealed that seeds of *Helicteres isora* did not germinate, while seeds of the remaining five species, *viz.*, *Adenanthera pavonina*, *Murraya paniculata*, *Psoralea corylifolia*, *Uraria lagopodioides* (L.) Desv. and *U. picta* germinated.

Results of seed germination in *Adenanthera pavonina*, *Murraya paniculata* and *Psoralea corylifolia*, *Uraria lagopodioides* and *U. picta* are presented in Table 1. The minimum days taken for germination of seeds in *Adenanthera pavonina* are 12 and the suitable time for seed sowing is April when the germination rate is the highest. Seeds of *Murraya paniculata* required minimum 36 days to germinate and the germination rate is found to be higher in April. In *Psoralea corylifolia* seeds were sown in different months but the highest percentage of seed germination was noted in July and the best time for seed sowing for this species is June. The minimum days taken to germinate the seeds were 10.

The present study reveals that in *Uraria picta* seeds sown after collection in December (12.12.2011) did not germinate, whereas seeds sown in mid of April (15.4.12) took 14 days indicating the minimum time for its germination. It is evident that in *Uraria lagopodioides* seeds sown after collection in December (12.12.2011) were not germinated as well. The minimum days required for seed germination in this species is 39 when seeds sown near mid April (Table 1).

Table 1. Results of seed germination of five species of medicinal plants.

Species	Date of seed collection	Date of seed sowing	No. of seeds sown	No. of seeds germinated	Days taken to germinate	% of germination
		4.3.2011	10	4	42-45	40%
Adenanthera pavonina L.	19.12.2010	25.3.2012	10	6	15	60%
		15.4.2012	10	10	12	100%
		1.2.2012	10	2	42	20%
Murraya paniculata (L.) Jack	29.1.2012	27.2.2012	10	2	36	20%
		13.3.2012	10	3	38	30%
		5.5.2012	10	2	12	20%
Psoralea corylifolia L.	5.5. 2012	19.6.2012	10	4	10-17	40%
		14.9.2012	10	2	13	20%
		12.12.2011	10	0	-	-
Uraria lagopodioides (L.) Desv.	29.11. 2011	1.1. 2012	10	0	-	-
		14.2.2012	10	1	80	10%
		12.4.2012	10	1	39	10%
		12.12.2011	10	0	-	-
U. picta (Jacq.) Desv. *ex* DC.	7.12. 2011	1.1. 2012	10	0	-	-
		27.2.2012	10	1	33	10%
		15.4.2012	10	1	14	10%

In *Helicteres isora*, seeds were not germinated indicating that they are not suitable for propagation through seeds. Consequently other mode of propagation like stem cutting was done for this species. Table 2 shows result of stem cutting experiment for *Helicteres isora*. The result indicates that *Helicteres isora* can be propagated by stem cutting. Therefore, propagation should be done by stem cutting method. The study also indicates that plant from stem cutting takes only 5-6 months to flowers and set fruits. The development of seedlings from seeds/ stem cutting up to maturity in the taxa studied is displayed in Plate 1.

Table 2. Result of stem cutting experiment in *Helicteres isora* L.

Date of stem cutting	Length of the stems (cm)	Date of leaf bud formation	Time taken to appear leaf bud (days)	Average time (days)	Date of flowering	Date of fruit formation
	25	18.5.2012	32		29.9.2012	22.10.2012
	25	18.5.2012	32		3.10.2012	1.11.2012
17.4.2012	25	22.5.2012	36	34.5	Died	Died
	25	25.5.2012	38		Died	Died

Plate 1. Development stages of six medicinal plants. A-D *Adenanthera pavonina* (A. seeds; B. seedling; C. mature plants with flowering stage; D. fruiting stage). E-H *Murraya paniculata* (E. seeds; F. seedling; G. mature plants with flowering stage; H. fruits). I-L *Psoralea corylifolia* (I. seeds; J. seedling; K. mature plants with flowering stage; L. fruits). M-P *Uraria lagopodioides* (M. seeds; N. seedling; O. mature plants with flowering stage; P. fruiting stage). Q-T *Uraria picta* (Q. seeds; R. seedling; S. mature plants with flowering stage; T. fruiting stage). U-X *Helicteres isora* (U. stem cuttings; V. initiation of leaves; W. flowering; X. fruiting stage).

In the present study we investigated seed germination of six medicinal plants. The present study reveals that epigeal germination is found in *Adenanthera pavonina*, *Murraya paniculata*, *Psoralea corylifolia*, *Uraria lagopodioides* and *U. picta,* whereas seeds of *Helicteres isora* failed to germinate. Time taken by the seeds to germinate varies from 10 days in *Psoralea corylifolia* to 80 days in *Uraria lagopodioides* (Table 1). Important factors controlling the variation in seed dormancy within species include the environment of the mother plant during the time of seed maturation and environmental conditions (Liebst and Schneller, 2008). Certain environmental conditions may be required to break dormancy, and other conditions are often required to permit germination after dormancy is broken (Foley, 2001). Seeds of many species require days, weeks, or months at low temperatures to break dormancy (Bewley and Black, 1994; Vleeshouwers *et al.,* 1995), whereas others require warm temperatures for after-ripening to germinate when permissive conditions arrive (Baskin and Baskin, 1972). In the present study it required around two weeks to break the seed dormancy in *Psoralea corylifolia*, whereas, in *Uraria lagopodioides* it took one to three months to break the dormancy.

The environmental factors that could affect seed dormancy are time of seed harvest, length of seed storage, relative humidity and photoperiod (Baskin and Baskin, 1973). In this study seeds of different taxa were not collected at the same time because of the differences in the period of seed production among the taxa. Therefore, the level of dormancy observed may be affected by environmental factors. The level of dormancy observed may be affected by environmental factors. The number of days for germination is related to the size of seeds, the largest seeds germinated faster than the smaller seeds (Gerry and Wilson, 1995). However, our results were found incongruent with Gerry and Wilson (1995). Since different environmental factors affect on seed germination therefore it is necessary to carry out a detailed study considering the factors that might through more light on germination patterns which are considered to be of taxonomic importance (Vogel, 1980).

References

Baskin, J.M. and Baskin, C.C. 1972. Ecological life cycle and physiological ecology of seed germination of Arabidopsis thaliana. Can. J. Bot. **50**: 353-360.

Baskin, J.M. and Baskin, C.C. 1973. Plant population differences in dormancy and germination characteristics of seeds: heredity or environment? Am. Midl. Nat. **90**: 493-498.

Beweley, J.D. and Black, M. 1994. Dormancy and the control of germination. Seeds: physiology of development and germination. 2nd ed. Plenum, New York.

Chauhan, B.S. and Johnson, D.E. 2008. Influence of environmental factors on seed germination and seedling emergence of *Eclipta* (*Eclipta prostrata*) in a tropical environment. Weed Sci. **56**: 383-388.

Clements, C.D., Harmon, D. and Young, J.A. 2010. Diffuse knapweed (*Centaurea diffusa*) seed germination. Weed Science **58**: 369-373.

de Padua, L.S., Bunyapraphatsara, N. and Lemmens, R.H.M.J., 1999. Plant resources of south-east Asia, No. **12**(1). Medicinal and poisonous plants 1. Backhuys Publishers, Leiden, the Netherlands. 711 pp.

Foley, M.E. 2001. Seed dormancy: an update on terminology, physiological genetics, and quantitative trait loci regulating germinability. Weed Sci. **49**: 305-317.

Gerry, A.K. and Wilson, S.D. 1995. The influence of initial size on the competitive responses of six plant species. Ecology **76**: 272-279.

Ghani, A. 2003. Medicinal Plants of Bangladesh with Chemical Constituents and Uses (Second Edition). Asiatic Society of Bangladesh, Dhaka. 603 pp.

Han, C.-Y. and Long, C.-L. 2010. Seed dormancy, germination and storage behavior of *Magnolia wilsonii* (Magnoliaceae), an endangered plant in China. Acta Bot. Yun. **32**(1): 47-52.

Hassan, M.A. and Fardous, Z. 2003. Seed germination, pollination and phenology of *Gloriosa superba* L. (Liliaceae). Bangladesh J. Plant Taxon. **10**(1): 95-97.

Kameneva, L.A. and Koksheeva, I.M. 2013. Reproductive biology of seven taxa of *Magnolia* L. in the south of Russian Far East. Bangladesh J. Plant Taxon. **20**(2): 163-170.

Liebst, B. and Schneller, J.S. 2008. Seed dormancy and germination behavior in two *Euphrasia* species (Orobanchaceae) occurring in the Swiss Alps. Bot. J. Linn. Soc. **156**: 649-656.

Liza, S.A., Rahman, M.O., Uddin, M.Z., Hassan, M.A. and Begum, M. 2010. Reproductive biology of three medicinal plants. Bangladesh J. Plant Taxon. **17**(1): 69-78.

Mayer, A.M. and Poljakoff-Mayber, A. 1989. The germination of seeds. Pergamon Press, New York, NY.

Rahman, M.Z., Rahman, M.O. and Hassan, M.A. 2012. Seed germination of two medicinal plants: *Desmodium pulchellum* (L.) Benth. and *D. triflorum* (L.) DC. Bangladesh J. Plant Taxon. **19**(2): 209-212.

Yusuf, M., Chowdhury, J.U., Haque, M.N. and Begum, J., 2009. Medicinal Plants of Bangladesh. Bangladesh Council of Scientific and Industrial Research, Chittagong, Bangladesh.

Vandelook, F. and van Assche, J.A. 2009. Temperature conditions control embryo growth and seed germination of *Corydalis solida* (L.) Clairv., a temperate forest spring geophyte. Plant Biology **11**: 899-906.

Vleeshouwers, L.M., Bouwmeester, H.J. and Karssen, C.M. 1995. Redefining seed dormancy: an attempt to integrate physiology and ecology. J. Ecol. **83**:1031-1037.

Vogel, E.F. 1980. Morphological types in dicot seedlings with reference to their origin. Bulletin de la Societe Botunique de France **126**(3): 173-182.

3

GENETIC DIVERSITY AND INTERSPECIFIC RELATIONSHIPS OF SOME *ALLIUM* L. SPECIES USING INTER SIMPLE SEQUENCE REPEAT MARKERS

LEILA SAMIEI[1], MAHNAZ KIANI, HOMA ZARGHAMI, FARSHID MEMARIANI[2] AND MOHAMMAD REZA JOHARCHI[2]

Department of Ornamental Plants, Research Center for Plant Sciences, Ferdowsi University of Mashhad, Mashhad, Iran

Keywords: *Allium* L.; Genetic relationship; ISSR; Molecular marker.

Abstract

In this study genetic diversity and interspecific relationships of 11 *Allium* L. species from Khorassan province of Iran including 32 accessions were investigated by inter simple sequence repeat (ISSR) markers. Nine ISSR primers produced a total of 80 polymorphic markers and revealed high polymorphism among the studied species. The average gene diversity, effective number of alleles and Shannon's information index were 0.2, 1.28 and 0.3, respectively. *Allium kuhsorkhense* exhibited the greatest level of variation (H_e: 0.18), whereas *A. stipitatum* demonstrated the lowest level of variability (H_e: 0.05). UPGMA (Unweighted Pair Group Method with Arithmetic mean) analysis showed that *Allium* accessions have a similarity range of 0.60 to 0.95. *Allium scapriscapum* composed the most distant group in the dendrogram. The clustered groups of *Allium* species clearly reflect the recent taxonomic concept of the genus at the subgenus and section levels. The present study showed that the ISSR technique is an effective molecular approach for analyzing genetic diversity and relationship in *Allium* species.

Introduction

The genus *Allium* L. is a member of Amaryllidaceae (APG III, 2009), subfamily Allioideae, tribe Allieae (Chase *et al.*, 2009; Reveal and Chase, 2011). It is one of the largest genera of monocots and comprises more than 900 species naturally occurring in the Northern Hemisphere (Fritsch and Abbasi, 2013). This genus has a main centre of diversity in the eastern Mediterranean area as well as southwest and central Asia (Fritsch and Friesen, 2002). *Allium* is a typical genus for Irano-Turanian floristic region and displays a high level of specific endemism there (Matin, 1992). There are nearly 50 *Allium* species, which are cultivated widely in the world and many more wild species are utilized locally for human consumption as spices, vegetables, medicinal and ornamental plants (Friesen *et al.*, 2006).

Allium consists of perennial herbs mostly characterized by tunicate bulbs, narrow basal leaves, umbellate or head-like inflorescences, flowers with 6 free or almost free tepals, and an onion-like odour and taste due to the presence of cystine sulphoxides (Li *et al.*, 2010). Many studies assessing morphological and anatomical characters of *Allium* species have been performed and numerous data have so far been published (Friesen, 1995; Mathew, 1996; Fritsch and Friesen, 2002; Kovtonyuk *et al.*, 2009). However, due to the close morphological similarities of the species, over reliance on dried specimens, and high degree of polymorphism of specific morphological traits (Khassanov and Fritsch, 1994; Mes *et al.*, 1997), many gaps still remain in

---

[1]Corresponding author. Email: samiei@um.ac.ir
[2]Department of Botany, Research Center for Plant Sciences, Ferdowsi University of Mashhad, Mashhad, Iran.

our knowledge of infrageneric taxonomy and differentiation and evolution in the genus (Rabinowitch and Brewster, 1990; Rabinowitch and Currah, 2002).

DNA-based molecular markers have been used previously in the studies of genetic diversity and phylogenetic analysis of *Allium* (Mes *et al.*, 1999; Friesen *et al.*, 2006; Gurushidze *et al.*, 2008; Mukherjee *et al.*, 2013). Of the different molecular markers, inter sample sequence repeat (ISSR) marker has been widely used to access species genetic diversity and relationships because of its cost effectiveness, simple operation as well as the need of very little starting DNA template (Lin *et al.*, 2009; Uysal *et al.*, 2010). In addition, previous studies of evaluating the phylogenetic relationship of Korean *Allium* species using ISSR marker indicated that these markers were highly informative in *Allium* (Hao *et al.*, 2002).

Recent advances in taxonomy and classification of *Allium* have shown that, there are about 135 species of *Allium* including 7 subgenera and 32 sections in Iran (Fritsch and Maroofi, 2010; Memariani *et al.*, 2012; Fritsch and Abbasi, 2013). North-eastern part of Iran (Khorassan provinces) with about 35 species is one of the most important centres of diversity of genus *Allium* in the country (Memariani *et al.*, 2007). Previously there have been some studies on the taxonomy of *Allium* (Fritsch *et al.*, 2006; Fritsch and Abbasi, 2013), however, except for a few studies focusing on the diversity of one species (Abdoli *et al.*, 2009; Ebrahimi *et al.*, 2009), there has not been any reports corresponding to molecular genetic diversity and genetic relationship of *Allium* in Iran. The present study was designed to explore the genetic diversity and interspecific relationships of some *Allium* species in north-east Iran and to evaluate the potential of ISSR marker in detecting the genetic variability of native Alliums.

Materials and Methods
Plant materials:

A total of 32 accessions representing 11 species of *Allium* were collected from North Khorassan and Razavi Khorassan provinces, located in northeast of Iran during 2012–2013 (Table 1). The samples were identified based on morphological characteristics and diagnostic descriptions of the species in the relevant literature (Wendelbo, 1971; Fritsch and Abbasi, 2013). Modern concepts of infrageneric classification of the genus are based on Friesen *et al.* (2006), Fritsch *et al.* (2010), and Fritsch and Abbasi (2013).

DNA isolation:

Total genomic DNA was extracted based on CTAB method (Doyle and Doyle, 1990) using Accuprep genomic DNA extraction kit (Bioneer, Korea) following manufacturer's instructions. The relative purity and concentration of extracted DNA was estimated with spectroscopy and Lambda DNA (Thermo scientific, USA) using a known concentration as a reference.

ISSR amplification:

A set of 20 ISSR primers (University of British Colombia, Canada) was screened to generate the molecular profiles. Nine out of 20 primers were selected because of their consistent amplification and clear banding pattern. The primers sequences are listed in Table 2. PCR condition was optimized using different concentration of template DNA and Mg as well as different annealing temperature. PCR was done with 10 ng template genomic DNA, 5 µl of Taq DNA Polymerase, 2× Master Mix RED (Ampliqon, Denmark), 1.5 mM $MgCl_2$, 0.3 µM primer, in a total volume of 10 µL. DNA amplification was performed on Ependorf Master cycler gradient (Ependorf Scientific, Germany) using the following condition: an initial denaturation step of 94˚C for 5 min followed by 37 cycles of 94˚C for 25 s, optimized annealing temperature for 25 s, 72˚C for 1 min and a final extension at 72˚C for 5 min. The amplification products were separated by electrophoresis on 1.5% (W/V) agarose gel in 0.5× tris-borate-ethylenediamin tetra acetic acid

Table 1. List of species of *Allium* L. employed in the present study.

Accession code	Species	Section	Subgenus	Location
1	*Allium ampeloprasum* L.	*Allium*	*Allium*	North Khorassan: Rein
2	*A. ampeloprasum* L.	*Allium*	*Allium*	Razavi Khorassan: Ferizi
3	*A. ampeloprasum* L.	*Allium*	*Allium*	Razavi Khorassan: Andishish village
4	*A. ampeloprasum* L.	*Allium*	*Allium*	Razavi Khorassan: Ferizi
5	*A. atroviolaceum* Boiss.	*Allium*	*Allium*	Razavi Khorassan: Ferizi
6	*A. atroviolaceum* Boiss.	*Allium*	*Allium*	Razavi Khorassan: Ferizi
7	*A. atroviolaceum* Boiss.	*Allium*	*Allium*	Razavi Khorassan: Ferdowsi Univ. campus
8	*A. umbilicatum* Boiss.	*Avulsea*	*Allium*	Razavi Khorassan: Tandooreh
9	*A. umbilicatum* Boiss.	*Avulsea*	*Allium*	Razavi Khorassan: Tandooreh
10	*A. umbilicatum* Boiss.	*Avulsea*	*Allium*	Razavi Khorassan: Akhlamad
11	*A. cristophii* Trautv.	*Asteroprason*	*Melanocrommyum*	Razavi Khorassan: Tandooreh
12	*A. cristophii* Trautv.	*Asteroprason*	*Melanocrommyum*	Razavi Khorassan: Akhlamad
13	*A. cristophii* Trautv.	*Asteroprason*	*Melanocrommyum*	Razavi Khorassan: Bazangan
14	*A. ellisii* Hook. f.	*Asteroprason*	*Melanocrommyum*	Razavi Khorassan: Hezar Masjed Mountain
15	*A. altissimum* Regel	*Procerallium*	*Melanocrommyum*	Razavi Khorassan: Tandooreh
16	*A. altissimum* Regel	*Procerallium*	*Melanocrommyum*	Razavi Khorassan: Shamkhal
17	*A. altissimum* Regel	*Procerallium*	*Melanocrommyum*	North Khorassan: Zu-e Eram
18	*A. stipitatum* Regel	*Procerallium*	*Melanocrommyum*	Razavi Khorassan: Ferizi
19	*A. stipitatum* Regel	*Procerallium*	*Melanocrommyum*	Razavi Khorassan: Ferizi
20	*A. stipitatum* Regel	*Procerallium*	*Melanocrommyum*	Razavi Khorassan: Ferizi
21	*A. sarawschanicum* Regel	*Megaloprason*	*Melanocrommyum*	Razavi Khorassan: Shamkhal
22	*A. sarawschanicum* Regel	*Megaloprason*	*Melanocrommyum*	Razavi Khorassan: Shamkhal
23	*A. kuhsorkhense* R.M. Fritsch & Joharchi	*Asteroprason*	*Melanocrommyum*	Razavi Khorassan: Ferizi
24	*A. kuhsorkhense* R.M. Fritsch & Joharchi	*Asteroprason*	*Melanocrommyum*	Razavi Khorassan: Ferizi
25	*A. kuhsorkhense* R.M. Fritsch & Joharchi	*Asteroprason*	*Melanocrommyum*	Razavi Khorassan: Ferizi
26	*A. kuhsorkhense* R.M. Fritsch & Joharchi	*Asteroprason*	*Melanocrommyum*	Razavi Khorassan: Torbat-e-Jam
27	*A. kuhsorkhense* R.M. Fritsch & Joharchi	*Asteroprason*	*Melanocrommyum*	Razavi Khorassan: Torogh
28	*A. scabriscapum* Boiss.	*Scapriscapa*	*Reticulatobulbosa*	Razavi Khorassan: Tandooreh
29	*A. scabriscapum* Boiss.	*Scapriscapa*	*Reticulatobulbosa*	Razavi Khorassan: Allah-o Akbar Mountain
30	*A. oschaninii* O. Fedtsch.	*Cepa*	*Cepa*	Razavi Khorassan: Ferizi
31	*A. oschaninii* O. Fedtsch.	*Cepa*	*Cepa*	Razavi Khorassan: Ferizi
32	*A. oschaninii* O. Fedtsch.	*Cepa*	*Cepa*	Razavi Khorassan: Akhlamad

(TBE) buffer at 90 V, stained with DNA green viewer (Pars Tous, Iran) and visualized under ultraviolet (UV) light in gel documentation system (UVI doc, UK). A 100 bp DNA ladder (Thermo scientific, USA) was used as molecular size standard. PCR amplification was repeated twice or sometimes more for each primer to ensure the reproducibility of the results.

Table 2. Characteristics of ISSR markers and genetic diversity statistics.

Primers	Sequence	N	A_e	H_e	I
UBC 807	AGA GAG AGA GAG AGA GT	9	1.34 (0.182)	0.24 (0.102)	0.40 (0.132)
UBC 808	AGA GAG AGA GAG AGA GC	7	1.23 (0.103)	0.18 (0.068)	0.32 (0.095)
UBC 809	AGA GAG AGA GAG AGA GG	8	1.20 (0.136)	0.16 (0.085)	0.29 (0.118)
UBC 811	GAG AGA GAG AGA GAG AC	7	1.21 (0.091)	0.17 (0.061)	0.31 (0.085)
UBC 827	ACA CAC ACA CAC ACA CG	9	1.16 (0.079)	0.13 (0.057)	0.25 (0.084)
UBC 834	AGA GAG AGA GAG AGA GYT	11	1.26 (0.154)	0.19 (0.097)	0.33 (0.134)
UBC 835	AGA GAG AGA GAG AGA GYC	7	1.60 (0.365)	0.35 (0.160)	0.52 (0.193)
UBC 840	GAG AGA GAG AGA GAG AYT	6	1.68 (0.268)	0.39 (0.122)	0.57 (0.146)
UBC 855	ACA CAC ACA CAC ACA CYT	16	1.15 (0.061)	0.13 (0.047)	0.25 (0.074)
Average		8.88	1.28	0.2	0.34

N = number of bands; A_e = number of effective alleles; H_e = expected heterozigosity; I = Shannon's information index; Standard errors are in parentheses.

Data analysis:

The reproducible and well resolved fragments obtained from ISSR analysis were scored as binary code, *viz*. presence (1) and absence (0) of homologous bands. The binary data matrix was analyzed using NTSYS-PC version 2.1 software package (Rohlf, 2000). The pairwise genetic distances among all accessions was calculated based on Nei (1978) similarity coefficient. Genetic diversity (H_e) was calculated per primer and for each species (except for *Allium ellisii* Hook. f. for which only one accession was available) using POPGENE software, version 1.32 (Yeh and Boyle, 1997). A dendrogram was constructed by using the unweighted pair group method with arithmetic mean (UPGMA) employing the SAHN (Sequential Agglomerative Hierarchical and Nested) module of NTSYS-PC to show a phenetic representation of genetic relationships as revealed by similarity coefficient. Percentage of polymorphic bands (PPB) was calculated by dividing the number of polymorphic bands by total number of bands surveyed.

Results

Nine ISSR primers generated 80 bands corresponding to an average of 8.8 bands per primer (Table 2). The fragment size varied from 100 to 2200 bp and the number of bands ranged from 6 (UBC 840) to 16 (UBC 855). All of the 80 bands detected by ISSR primers were polymorphic among the individuals, *i.e.* the percentage of polymorphic bands was 100% for each primer. Considering all accessions, the average gene diversity, effective number of alleles and Shannon's information index was 0.2, 1.28 and 0.3, respectively. The genetic diversity generated by each primer varied from 0.38 (primer UBC 840) to 0.12 (UBC 855). The average effective number of alleles and Shannon's information index was 1.28 and 0.34, respectively. Among the 11 species, *A. kuhsorkhense* R.M. Fritsch & Joharchi and *A. ampeloprasum* L. exhibited the highest variability (PPB: 42.5% and 37.5%, H_e: 0.18 and 0.14, respectively) whereas the species *A. stipitatum* Regel and *A. sarawschanicum* Regel presented the least variability (PPB: 12.5% and 16.25%, and H_e: 0.05 and 0.07, respectively) as shown in Table 3.

A dendrogram generated based on Nei's genetic distances and UPGMA method revealed genetic relationships among *Allium* species and accessions (Fig. 1). The high cophenetic correlation (r = 0.95) obtained indicating a good fit between the dendrogram clusters and the distance matrix. The dendrogram displayed four main groups corresponding to four subgenera: *Allium*, *Melanocrommyum* (Webb & Berthel.) Rouy, *Cepa* (Mill.) Radić, and *Reticulatobulbosa* (Kamelin) N. Friesen. The most distant group comprised two accessions of *A. scabriscapum* with low genetic similarity coefficient (Nei = 0.22) belonging to the subgenus *Reticulatobulbosa*. The only species present in section *Cepa* in this study was *A. oschaninii* B. Fedtsch. which formed a distinct group. The largest group corresponds to the subgenus *Melanocrommyum*. It, however, is divided into three subclusters each composing the species of the same section: section *Procerallium* comprises the accessions of *A. altissimum* and *A. stipitatum*; *A. sarawshanicum* Regel (section *Megaloprason* Wendelbo) makes a separate cluster; and *A. cristophii* Trautv. and *A. ellisii* comprise the section *Asteroprason* R.M. Fritsch (subsection *Christophiana* Tscholok.). Five accessions of *A. kuhsorkhense* with the highest genetic diversity (0.18) placed in section *Asteroprason* (subsection *Asteroprason*). Species belonging to the subgenus *Allium* formed a separate cluster. This cluster, however, is divided into two subclusters correspond to sections *Allium* and *Avulsea* F.O. Khassanov.

Table 3. Diversity parameters of *Allium* species.

Species	Sample size	H_e	PPB (%)
A. altissimum Regel	3	0.1237	35
A. ampeloprasum L.	4	0.1445	37.5
A. atroviolaceum Bioss.	3	0.0802	21.25
A. cristophii Trautv.	3	0.0778	21.25
A. kuhsorkhense R.M. Fritsch & Joharchi	5	0.1779	42.5
A. oschaninii O. Fedtsch	3	0.0904	26.25
A. sarawschanicum Regel	2	0.0673	16.25
A. scabriscapum Bioss.	2	0.0725	17.5
A. stipitatum Regel	3	0.0516	12.5
A. umbilicatum Bioss.	3	0.1102	25

H_e = expected heterozygosity; PPB = percentage of polymorphic bands.

Discussion

In the present study, nine ISSR primers yielded a total of 80 reproducibale bands with an average of 8.8 bands per primer, which was higher than in some other studies that have applied ISSR markers to *Allium* species. For example, in an analysis of 24 accessions of 13 *Allium* species, Son *et al.* (2012) detected 3 to 11 alleles per locus (average 7.5 alleles per primer) using 20 ISSR markers. In our study, a very high level of ISSR polymorphisms was detected in the *Allium* species indicating ISSR-PCR as a reliable technique for fingerprinting in the genus. ISSR markers have been widely employed in assessment of genetic relationships within and between plant species (Thul *et al.*, 2012; Liu *et al.*, 2013). Although there are not many reports on application of ISSR markers for analyzing the genetic relationships among *Allium*, the efficacy of ISSR markers on revealing the classification of *Allium* species has been strongly supported by previous studies (Hao *et al.*, 2002; Son *et al.*, 2012). Furthermore, the homology of ISSR bands between *Allium* species has been formerly confirmed by sequence analysis (Son *et al.*, 2012).

In this study, the genetic diversity of the endemic species *A. kuhsorkhense*, as well as the genetic relationships among *Allium* species were studied for the first time. The study revealed that *A. kuhsorkhense* is the most diverse species among the species employed in Khorassan. The accessions of *A. kuhsorkhense* were collected from nearly distant area. This endemic species has the widest distribution range among the other species of the endemic *Allium* section *Asteroprason* (Memariani *et al.*, 2012). Hamrick (1989) stated that the wide distribution of a species can increase the rate of genetic diversity among the accessions. The clustered groups of *Allium* species clearly reflect the recent taxonomic concept of the genus at the subgenus and section levels. Subgenus *Melanocrommyum* comprises the largest group in the dendrogram. Three sections within the subgenus were detected and clearly identified, which is in consistent with previous studies (Gurushidze *et al.*, 2008; Fritsch *et al.*, 2010). In section *Asteroprason*, two main subclusters support its morphological classification into two subsections *Christophiana* and *Asteroprason* (Fritsch and Maroofi, 2010; Memariani *et al.*, 2012).

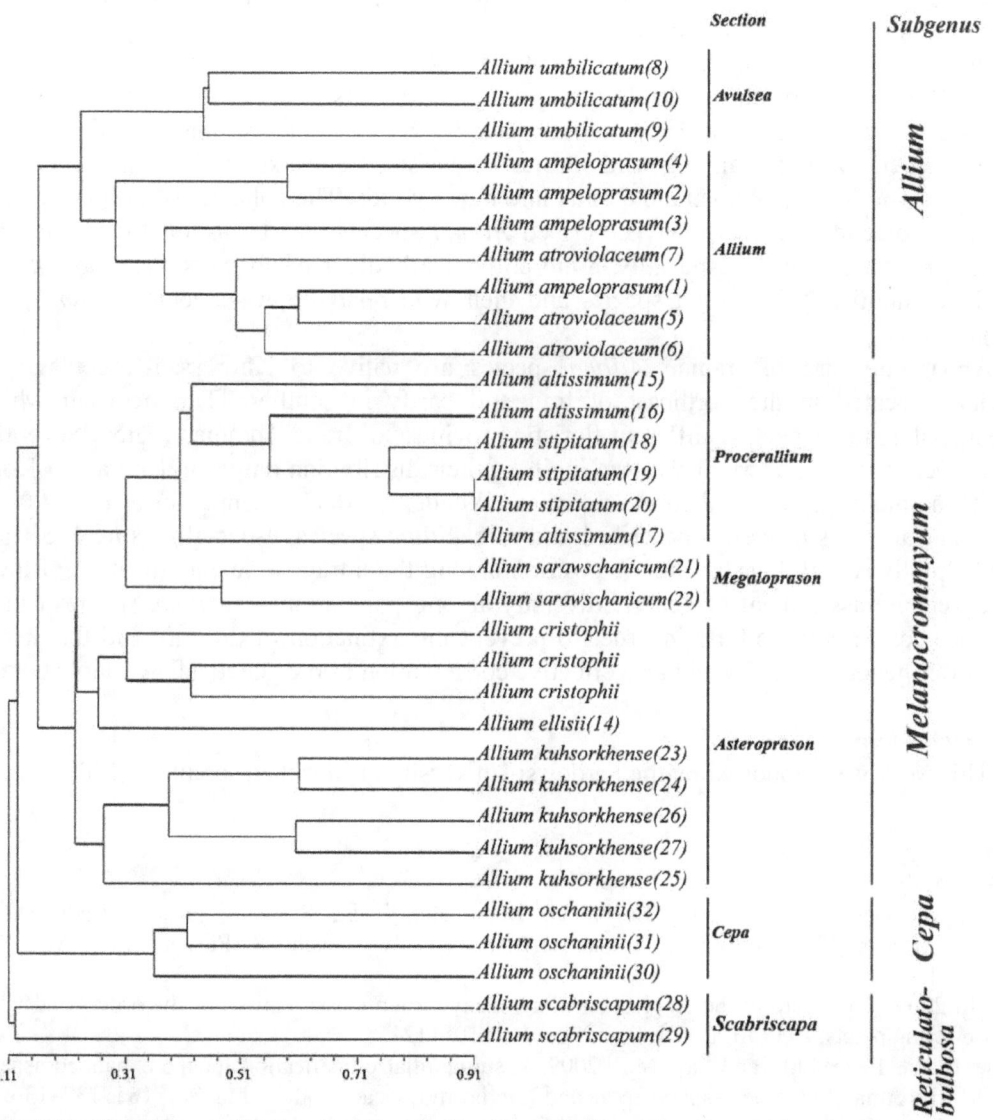

Fig. 1. UPGMA dendrogram showing species relationships of *Allium* based on Nei's genetic distance. The numbers in parentheses correspond accession codes of Table 1.

Within all sections of subgenus *Melanocrommyum*, the species were clearly distinguished except for *Procerallium* R.M. Fritsch. In this section, the accessions of *A. stipitatum* positioned among the accessions of *A. altissimum*. Indeed these two species are morphologically very similar and difficult to be distinguished properly. However, it is accepted that *A. altissimum* is more slender and somewhat smaller than typical *A. stipitatum* possessing narrower and glabrous (or at the most only sparsely toothed) leaves, a smaller umbel, and more intensely coloured, sublinear, in late anthesis spirally enrolled tepals (Fritsch and Abbasi, 2013). Our ISSR analysis is in agreement with the molecular analyses based on ITS sequences of nuclear rDNA, sequences of the plastid *trn*L-*trn*F region in which the cultivated strain of *A. altissimum* positioned among many accessions of Central Asian *A. stipitatum* underlining a high genetic similarity (Fritsch and Maroofi, 2010; Gurushidze *et al.*, 2010). Based on our analysis, the accessions of *A. ampeloprasum* and *A. atroviolaceum* (section *Allium*) are similarly positioned among each other, however, are well-separated from the accessions of *A. umblicatum* Boiss. (section *Avulsea*). Hirschegger *et al.* (2010) found similar results for *A. ampeloprasum* group based on nuclear and chloroplast DNA sequences analyses.

ISSR markers were highly informative at the section level as well as at the species level in the genus *Allium*. The resulting dendrogram was found consistent with the modern taxonomic classification, confirming that ISSR marker data can be used for taxonomic studies in the genus *Allium*. Genetic variation among wild species may assist plant taxonomists, and also breeders in identifying and introducing valuable traits into new hybrids. The collection and inclusion of more accessions of endemic and newly described *Allium* species will be useful for confirming their infrageneric classification, especially in morphologically diverse subgenus *Melanocrommyum* and also in economically important species and their wild relatives in subgenus *Allium* (garlic and leek).

About one-third of Iranian *Allium* species are native to Khorassan-Kopetdagh floristic province, located in the northeast of Iran and partly in southern Turkmenistan which is a transitional zone connecting different floristic provinces of Irano-Turanian region. Several *Allium* species occur in the eastern or western limits of their distribution ranges in Khorassan-Kopetdagh as well as many narrow and local endemics (Wendelbo, 1971; Memariani *et al.*, 2007, 2012). Molecular analyses on newly described and rare *Allium* species, especially using ISSR markers, may helpfully reveal their taxonomic position among the infrageneric classification of the genus. Moreover, the assessment of genetic diversity among populations of *Allium* species can help to prioritize conservation efforts in order to prevent the extinction of the rare and threatened taxa with lower genetic diversity and also effective conservation of the genetically variable taxa.

Acknowledgement

This work was supported by the Ferdowsi University of Mashhad (grant no. 17812).

References

Abdoli, M., Habibi-Khaniani, B., Baghalian, K., Shahnazi, S., Rassouli, H. and Badi, H.N. 2009. Classification of Iranian garlic (*Allium sativum* L.) ecotypes using RAPD marker. J. Med. Plants **8**: 45–51.

APG III 2009. An update of the Angiosperm Phylogeny Group classification for the orders and families of flowering plants: APG III. Bot. J. Linn. Soc. **161**: 105–121.

Chase, M.W., Reveal J.L. and Fay, M.F. 2009. A subfamilial classification for the expanded asparagalean families Amaryllidaceae, Asparagaceae and Xanthorrhoeaceae. Bot. J. Linn. Soc. **161**: 132–136.

Doyle, J.J. and Doyle, J.L. 1990. A rapid total DNA preparation procedure for fresh plant tissue. Focus **12**: 13–15.

Ebrahimi, R., Zamani, Z. and Kashi, A. 2009. Genetic diversity evaluation of wild Persian shallot (*Allium hirtifolium* Boiss.) using morphological and RAPD markers. Sci. Hortic. **119**: 345–351.

Friesen, N. 1995. The genus *Allium* L. in the flora of Mongolia. Feddes Repert. **106**: 59–81.

Friesen, N., Fritsch, R.M. and Blattner, F.R. 2006. Phylogeny and new intrageneric classification of *Allium* (Alliaceae) based on nuclear ribosomal DNA ITS sequences. Aliso **22**: 372–395.

Fritsch, R.M. and Abbasi, M. (Eds) 2013. A taxonomic review of *Allium* subg *Melanocrommyum* in Iran. Halberstädter Druckhaus Gmb H, Gatersleben, Germany, 240 pp.

Fritsch. R.M. and Friesen, N. 2002. Evolution, domestication and taxonomy. *In*: Rabinowitch, H.D. and Currah, L. (Eds), Allium Crop Science: Recent Advances. CABI Publishing, Wallingford, UK, pp. 5–30.

Fritsch, R.M. and Maroofi, H. 2010. New species and new records of *Allium* L. (Alliaceae) from Iran. Phyton **50**: 1–26.

Fritsch, R.M., Blattner, F.R. and Gurushidze, M. 2010. New classification of *Allium* L. subg *Melanocrommyum* (Webb & Berthel) Rouy (Alliaceae) based on molecular and morphological characters. Phyton **49**: 145–220.

Gurushidze, M., Fritsch, R.M. and Blattner, F.R. 2008. Phylogenetic analysis of *Allium* subg *Melanocrommyum* infers cryptic species and demands a new sectional classification. Mol. Phylogenet. Evol. **49**: 997–1007.

Gurushidze, M., Fritsch, R.M. and Blattner, F.R. 2010. Species-level phylogeny of *Allium* subgenus *Melanocrommyum*: Incomplete lineage sorting hybridization and *trn*F gene duplication. Taxon **59**: 829–840.

Hamrick, J.L. 1989. Isozymes and the analysis of genetic structure in plant populations. *In*: Soltis, D.E. and Soltis, P.S. (Eds), Isozymes in Plant Biology. Dioscorides Press, Portland, Oregon, USA, pp. 87–105.

Hirschegger, P., Jakse, J., Trontelj, P. and Bohanec, B. 2010. Origins of *Allium ampeloprasum* horticultural groups and a molecular phylogeny of the section *Allium* (*Allium*: Alliaceae). Mol. Phylogenet. Evol. **54**: 488–497.

Hao, G., Lee, D.H., Lee, J.S. and Lee, N.S. 2002. A study of taxonomical relationships among species of Korean *Allium* sect *Sacculiferum* (Alliaceae) and related species using inter-simple sequence repeat (ISSR) markers. Bot. Bull. Acad. Sin. **43**: 63–68.

Khassanov, F.O. and Fritsch, R.M. 1994. New taxa in *Allium* L. subgen. *Melanocrommyum* (Webb & Berth.) Rouy from Central Asia. Linzer Biologische Beitra **26**: 965–990.

Kovtonyuk, N.K., Barkalov, V.J.U. and Friesen, N. 2009. Synopsis of the family Alliaceae Borkh. (onions) of Asian parts of Russia. Turczaninowia **12**: 31–39.

Li, Q.Q., Zhou, S.D., He, X.J., Yu, Y., Zhang, Y.C. and Wei, X.Q. 2010. Phylogeny and biogeography of *Allium* (Amaryllidaceae: Allieae) based on nuclear ribosomal internal transcribed spacer and chloroplast rps16 sequences focusing on the inclusion of species endemic to China. Ann. Bot. **106**: 709–733.

Lin, X.C., Ruan, X.S., Lou, Y.F., Guo, X.Q. and Fang, W. 2009. Genetic similarity among cultivars of *Phyllostachys pubescens*. Plant Syst. Evol. **277**: 67–73.

Liu, Y., Zhang, J.M., Wang, X.G., Liu, F. and Shen, Z.B. 2013. Genetic diversity in *Vicia amoena* (Fabaceae) germplasm resource in China using SRAP and ISSR markers. Biochem. Syst. Ecol. **51**: 86–93.

Mathew, B. 1996. A review of *Allium* section Allium. Royal Botanic Gardens, Kew, UK, 176 pp.

Matin, F. 1992. The genus *Allium* in Iran, diversity, distribution and endemism. *In*: Hanelt, P., Hammer, K. and Knüpffer, H. (Eds), The genus *Allium* - Taxonomic Problems and Genetic Resources. Proc. Int. Symp. Gatersleben, June 11–13, 1991. IPK, Gatersleben, pp. 193–194.

Memariani, F., Joharchi, M.R. and Khassanov, F.O. 2007. *Allium* L. subgen. *Rhizirideum* sensu lato in Iran, two new records and a synopsis of taxonomy and phytogeography. Iran. J. Bot. **13**(1): 12–20.

Memariani, F., Joharchi, M.R. and Arjmandi, A.A. 2012. *Allium aladaghense* (Amaryllidaceae: Allieae) a new species of section *Asteroprason* from northeast of Iran. Phytotaxa **56**: 28–34.

Mes, T.H., Friesen, N., Fritsch, R.M., Klaas, M. and Bachmann, K. 1997. Criteria for sampling in *Allium* based on chloroplast DNA PCR-RFLPs. System. Bot. **22**: 701–712.

Mes, T.H, Fritsch, R.M., Pollner, S. and Bachmann, K. 1999. Evolution of the chloroplast genome and polymorphic ITS regions in *Allium* subg *Melanocrommyum*. Genome **42**: 237–247.

Mukherjee, A., Sikdar, B., Ghosh, B., Banerjee, A., Ghosh, E., Bhattacharya, M. and Roy, S.C. 2013. RAPD and ISSR analysis of some economically important species, varieties and cultivars of the genus *Allium* (Alliaceae). Turk. J. Bot. **37**: 605–618.

Nei, M. 1978. Estimation of average heterozygosity and genetic distance from a small number of individuals. Genetics **89**: 583–590.

Rabinowitch, H.D. and Brewster, J.L. 1990. Onions and allied crops. Vol. **1**. Botany, Physiology and Genetics. CRC Press, Boca Raton, FL, 288 pp.

Rabinowitch, H.D. and Currah, L. 2002. *Allium* Crop Science: Recent Advances. CAB International Wallingford, UK, 515 pp.

Reveal, J.L. and Chase, M.W. 2011. APG III: Bibliographical Information and Synonymy of Magnoliidae. Phytotaxa **19**: 71–134.

Rohlf, F.J. 2000. NTSYSpc: Numerical taxonomy and multivariate analysis system version 2.1. Exeter Publishing, Setauket, NY, 38 pp.

Son, J.H., Park, K.C., Lee, I., Kim, J.H. and Kim, N.S. 2012. Species relationships among *Allium* species by ISSR analysis. Hort. Environ. Biotechnol. **53**: 256–262.

Thul, S.T., Darokar, M.P., Shasany, A.K. and Khanuja, S.P. 2012. Molecular profiling for genetic variability in *Capsicum* species based on ISSR and RAPD markers. Mol. Biotechnol. **51**: 137–147.

Uysal, H., Fu, Y.B., Kurt, O., Peterson, G.W., Diederichsen, A. and Kusters, P. 2010. Genetic diversity of cultivated flax (*Linum usitatissimum* L.) and its wild progenitor pale flax (*Linum bienne* Mill.) as revealed by ISSR markers. Genet. Resour. Crop Evol. **57**: 1109–1119.

Wendelbo, P. 1971. Alliaceae. *In*: Rechinger, K.H. (Ed.), Flora Iranica, Vol. **76**. Akademische Druck-und Verlagsanstalt, Graz, Austria, 100 pp.

Yeh, F.C. and Boyle, T.J.B. 1997. Population genetic analysis of codominant and dominant markers and quantitative traits. Belg. J. Bot. **129**: 157–163.

4

PHYLOGENETIC IMPLICATION OF MOLECULAR GENOTYPING OF *EURYOPS JABERIANA* ABEDIN & CHAUDHARY (ASTERACEAE)

M. Ajmal Ali[1], Joongku Lee[2], M. Oliur Rahman[3], Fahad S.M. Al-Anazi, Fahad M.A. Al-Hemaid, A.A. Hatamleh, Changyoung Lee[4], B.J. Mylliemngap[5] AND A. Bhattacharjee[5]

Department of Botany and Microbiology, College of Science, King Saud University, Riyadh-11451, Saudi Arabia

Keywords: Euryops jaberiana; Asteraceae; nrDNA ITS; Genotyping; Saudi Arabia.

Abstract

The taxonomic status of *Euryops jaberiana* Abedin & Chaudhary (tribe Senecioneae, family Asteraceae), endemic to northern Saudi Arabia was evaluated based on molecular phylogenetic analyses of internal transcribed spacer sequence (ITS) of nuclear ribosomal DNA (nrDNA) in order to ascertain its position within the genus. The phylogenetic tree constructed by the Neighbour Joining, Maximum Parsimony and Maximum Likelihood analyses showed a clear resolution of taxon included in the analyses at the level of sections, and *E. jaberiana* nested within the clade of the section *Angustifoliae*. *E. jaberiana* showed proximity with the allied species *E. arabicus*; however, a total number of eight nucleotide differences were evident between *E. jaberiana* and *E. arabicus,* indicating *E. jaberiana* as distinct from its allied species.

Introduction

The genus *Euryops* (Cass.) Cass. belonging to the tribe *Senecioneae* of the family Asteraceae comprises approximately 100 species and displays a restricted distribution in Africa to Arabia and Socotra (Devos *et al.*, 2010). *Euryops* is characterized by perennial shrubs (except *E. annuus* Compt.), coriaceous leaves and yellow or orange-flowered capitula on simple peduncles, usually devoid of leaves or bracts. Despite the genus was divided into six sections *Angustifoliae, Brachypus, Chrysops, Euryops, Leptorrhiza* and *Psilosteum* based on morphology (Nordenstam, 1968), its phylogeny and phytogeography based on molecular data remains poorly understood (Nordenstam, 1969; Nordenstam *et al.*, 2009; Devos *et al.*, 2010). In Saudi Arabia, the genus *Euryops* is represented by only two species, *viz. E. arabicus* Steud. *ex* Jaub. & Spach, and *E. jaberiana* Abedin & Chaudhary. *E. arabicus* is the only species found outside of Africa and is endemic to Arabian Peninsula, while *E. jaberiana* is endemic to northern Saudi Arabia. Morphologically *E. jaberiana* very closely resembles with *E. arabicus* (Abedin and Chaudhary, 2000). Therefore, the main objectives of the present study are two-folds: i) to assess the

---

[1]Corresponding author. Email: majmalali@rediffmail.com
[2]Department of Environment and Forest Resources, Chungnam National University, 99 Daehak-ro, Yuseong-gu, Daejeon 34134, South Korea.
[3]Department of Botany, University of Dhaka, Dhaka 1000, Bangladesh.
[4]International Biological Material Research Center, Korea Research Institute of Bioscience and Biotechnology, 111 Gwahangno, Yuseong-gu, Daejeon 305 806, South Korea.
[5]Department of Biotechnology and Bioinformatics, North Eastern Hill University, Shillong 793002, Meghalaya, India.

phylogenetic relationships of *E. jaberiana* within the genus, and ii) to shed light on the molecular authentication of *E. jaberiana*.

Materials and Methods

Plant material:

Leaf materials of *Euryops jaberiana* were collected from the herbarium specimens [voucher-Saudi Arabia, Jabal Shaar near Al-Muwaylih, N. Hijaz, Alt. 1400-1500 m, 03 March 1988, S. Chaudhary and J. Thomas 16873, *Isotype*: (RIY)] housed at National Herbaium, Riyadh, Saudi Arabia (RIY).

Total genomic DNA extraction, amplification of ITS region and DNA sequencing:

The total genomic DNA was isolated using Qiagen DNeasy plant mini kit (Valencia, CA, USA). The internal transcribed spacer (ITS) sequences of nuclear ribosomal DNA (nrDNA) were amplified using forward primer ITS1 (5$'$-GTCCACTGAACCTTATCATTTAG-3$'$) and reverse primer ITS4 (5$'$-TCCTCCGCTTATTGATATGC-3$'$)] of White *et al.* (1990). The amplified product was sequenced on the ABI 3730 XL sequencing platforms by following methods described by Al-Hemaid *et al.* (2014) and Ali *et al.* (2015a).

Phylogenetic analysis:

The sequence of *E. jaberiana* (GenBank accession Number KU577443) was aligned with a total number of 17 representative sequences belongs to each section of the genus *Euryops* and an outgroup sequence of *Gymnodiscus capillaris* retrieved from GenBank (Table 1). The alignment was performed using CLUSTAL X version 1.81 (Thompson *et al.*, 1997). The alignment was manually adjusted using the software BioEdit (Hall, 1999). The Neighbour Joining (NJ) and also

Table 1. GenBank Accession number of Plant species used for molecular phylogentic analyses.

Group	Species	GenBank Acc. number
Ingroup	1. *Euryops annuus* Compt.	EU667487
	2. *Euryops anthemoides* B. Nord.	EU667501
	3. *Euryops arabicus* Steud.	EU667464
	4. *Euryops brachypodus* (DC.) B. Nord.	EU667485
	5. *Euryops brevilobus* Compt.	EU667488
	6. *Euryops dacrydioides* Oliv.	EU667529
	7. *Euryops decumbens* B. Nord.	EU667474
	8. *Euryops ericifolius* (Bel.) B. Nord.	EU667519
	9. *Euryops ericoides* (L.f.) B. Nord.	EU667509
	10. *Euryops evansii* Schltr.	EU667471
	11. *Euryops hypnoides* B. Nord.	EU667527
	12. *Euryops jaberiana* Abedin & Chaudhary	KU577443
	13. *Euryops montanus* Schltr.	EU667462
	14. *Euryops othonnoides* (DC.) B. Nord.	EU667503
	15. *Euryops pectinatus* (L.) Cass.	EU667514
	16. *Euryops pinifolius* A. Rich.	EU667530
	17. *Euryops speciosissimus* DC.	EU667717
	18. *Euryops trilobus* Harv.	EU667469
Outgroup	19. *Gymnodiscus capillaris* (L. f.) Less.	EU667515

the Maximum Parsimony (MP) and Maximum Likelihood (ML) analyses were carried out using PAUP (Swofford, 2002) and MEGA5 (Tamura *et al.*, 2011) respectively by the methods as described by Pandey and Ali (2012), Ali *et al.* (2013, 2015b), and Lee *et al.* (2013).

Results and Discussion

The present study revealed that the combined length of ITS region (ITS1-5.8S-ITS2) in *E. jaberiana* was 645 nucleotide base pair (bp). The ITS1 region was 260 bp (with GC content 43%), the 5.8S gene was 154 bp long (GC content 54%), and the ITS2 region was 231 bp (GC content 50%). The nrDNA in eukaryotes encodes for ribosome subunits, which occurs in thousands of copies (Prokopovich *et al.*, 2003) that simplify the amplification by polymerase chain reaction (PCR). The nrDNA consist of both highly variable parts of ITS region (i.e. ITS1 and ITS2) and the conserved 5.8S gene between ITS1 and ITS2 (Baldwin *et al.*, 1995). Although reliance on the use of ITS sequence of nrDNA as the sole source of phylogenetic evidence has come under serious criticism (Alvarez and Wendel, 2003); even then, it is one of the most common molecular markers used for generating species-specific phylogenetic inferences in most groups of plants, fungi and animals (Poczai and Hyvönen, 2010; Ali *et al.*, 2014) and DNA barcoding (Chen *et al.*, 2010; Yao *et al.*, 2010; Ali *et al.*, 2014, 2015c) owing to the patterns of polymorphism and ITS types which are specific to particular taxon and population (Baldwin *et al.*, 1995; Feliner *et al.*, 2004; Szabo *et al.*, 2005). The ITS sequence of nrDNA has gained much attention as smartest gene available for the genotyping of taxon and the epitome of species identification has thus now been changed due to application of genotyping in systematics (Ali *et al.*, 2013, 2014).

The BLAST search (Altschul *et al.*, 1990) of the generated nrDNA ITS sequence of *E. jaberiana* showed 99% identity with *E. arabicus*. The phylogenetic analyses revealed a total number of 610 positions in the final aligned dataset, of which 35 were parsimony informative. The MP analysis of the entire ITS region resulted in 82 maximally parsimonious trees (MPTs), the consistency index was 0.671, the retention index was 0.727, the composite index was 0.488 and homoplasy index 0.354. The phylogenetic tree recovered by the analyses provided a clear resolution of taxon at the section level which is consistent with previous study (Devos *et al.*, 2010).

Neighbour Joining (NJ) tree inferred from ITS sequence of nuclear ribosomal DNA of 18 species of *Euryops* revealed that *E. jaberiana* is phylogenetically most closely related to *E. arabicus* (Fig. 1). The NJ analysis recovered tree topology similar to MPT and MLT, and therefore, only the NJ topology with bootstrap support at the node is presented in Fig. 1.

The key morphological features which differentiate *E. jaberiana* from *E. arabicus* are: leaves 3-lobed at the tips, pappus hairs transparent or rarely dull white, and achenes glabrescent, while in *E. arabicus*, the leaves are unlobed, pappus hairs are dull white and achene densely lanate hairy (Abedin and Chaudhary, 2000). In both the MP and ML analyses, *E. jaberiana* nested within the clade of the section *Angustifoliae*. *E. jaberiana* shows proximity with *E. arabicus* (66% bootstrap support in MPT and 73% bootstrap support in MLT). A total of eight specific nucleotide differences *i.e.* at the alignment position 93 (A → T), 116 (G → C), 201 (T → C), 443 (C → G), 461 (T → G), 531 (T → C), 573 (C→T) and 611 (T→C) were detected between *E. jaberiana* and *E. arabicus* (Fig. 2). Thus on the basis of phylogenetic relationships of *E. jaberiana* within the genus and nucleotide differences, we herein recognized *E. jaberiana* as a distinct species and different from *E. arabicus*.

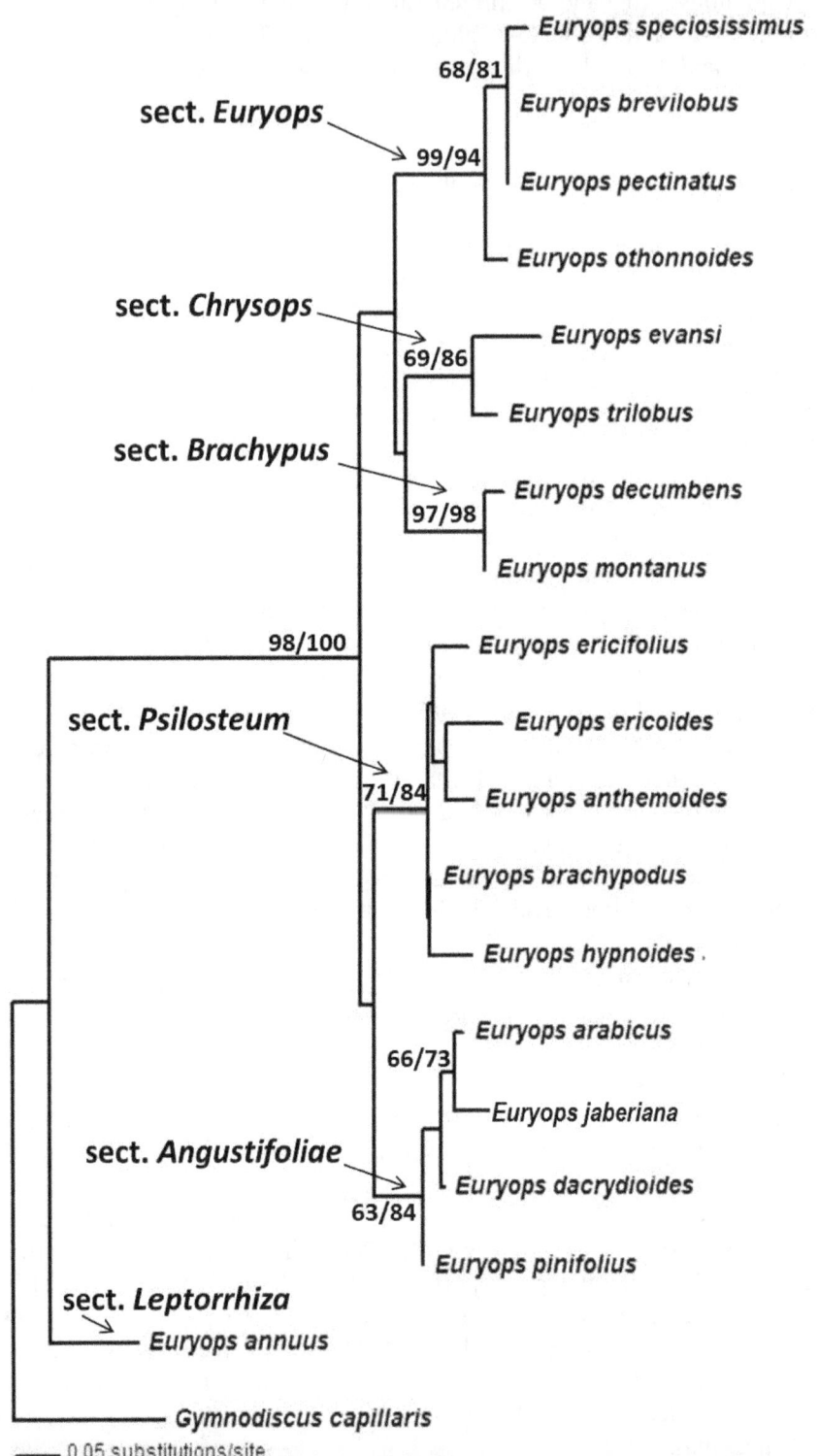

Fig. 1. The NJ tree inferred from Neighbour Joining analysis of ITS sequence of nuclear ribosomal DNA of 18 species of *Euryops*. The bootstrap (MP/ML) support greater than 50% in 1000 bootstrap replicates shown on the branch.

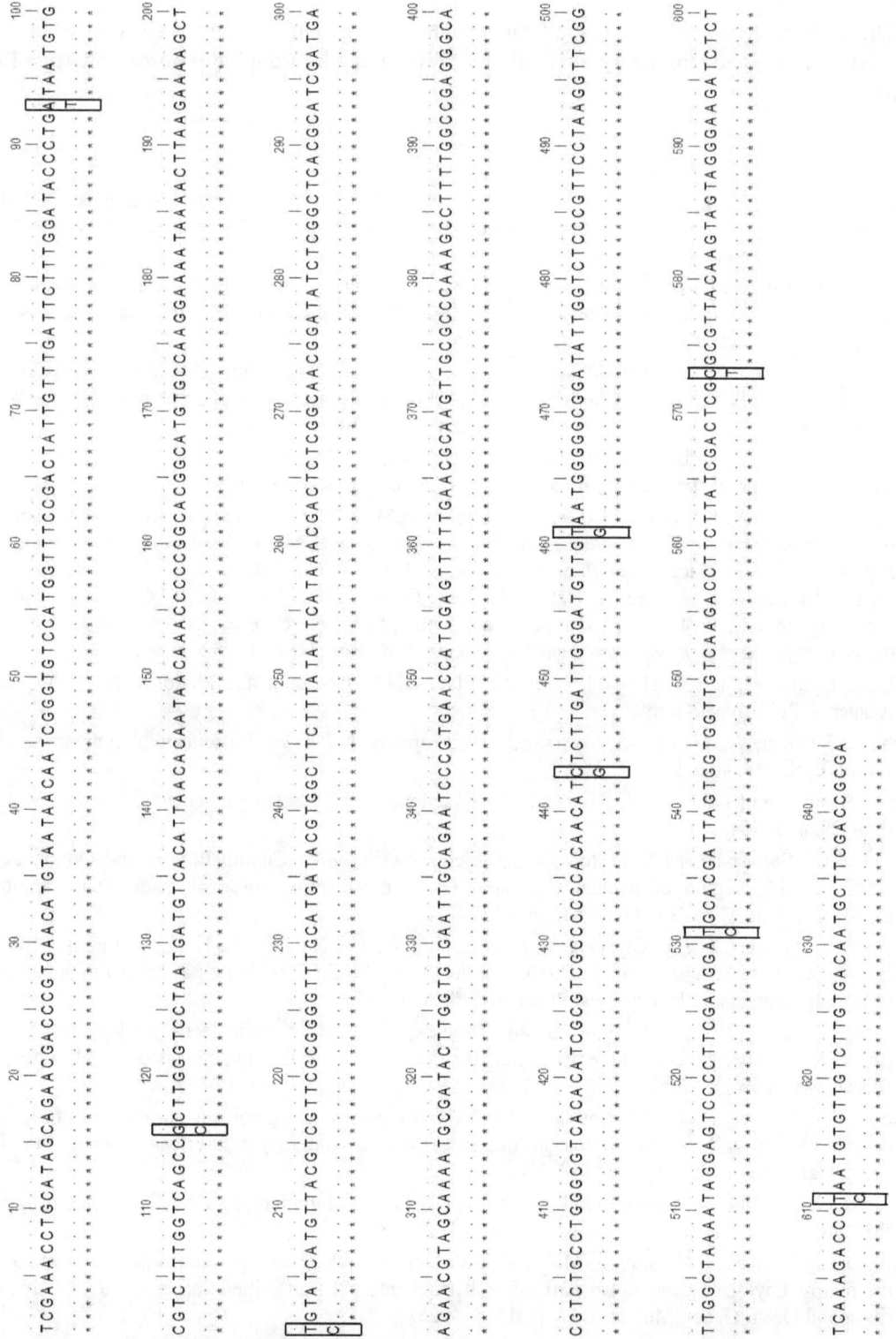

Fig. 2. Differences in the nucleotide base pairs position marked with box. Lane 1: *E. jabriana*, Lane 2: *E. arabicus*, and Lane 3: Clustal consensus.

Acknowledgement

The authors would like to extend their sincere appreciation to the Deanship of Scientific Research at King Saud University for funding of this research through the Research Group Project No. RGP-195.

References

Abedin, S. and Chaudhary, S. 2000. *Euryops*. *In*: Chaudhary, S. (Ed.), Flora of Saudi Arabia **II** (3): 191–192. Ministry of Agriculture and Water, National Herbarium, National Agriculture and Water Research Center, Riyadh, Saudi Arabia.

Al-Hemaid, F.M.A., Ali, M.A., Lee, J., Gyulai, G. and Pandey, A.K. 2014. Application of internal transcribed spacer of nuclear ribosomal DNA for identification of *Echinops mandavillei* Kit Tan. Bangladesh J. Plant Taxon. **21**(1): 33–42.

Ali, M.A., Al-Hemaid, F.M.A., Choudhary, R.K., Lee, J., Kim, S.Y. and Rub, M.A. 2013. Status of *Reseda pentagyna* Abdallah & A.G. Miller (Resedaceae) inferred from analysis of combined nuclear ribosomal and chloroplast sequence data. Bangladesh J. Plant Taxon. **20**(2): 233–238.

Ali, M.A., Gyulai, G., Norbert, H., Balázs, K., Al-Hemaid, F.M.A., Pandey, A.K. and Lee, J. 2014. The changing epitome of species identification - DNA barcoding. Saudi J. Biol. Sci. **21**(3): 204–231.

Ali, M.A., Lee, J., Kim, S.Y., Park, S.H. and Al-Hemaid, F.M.A. 2015a. Molecular phylogenetic analyses of internal transcribed spacer sequences of nuclear ribosomal DNA defined monophyly of the genus *Phytolacca* L. (Phytolaccaceae). Bangladesh J. Plant Taxon. **22**(1): 1–8.

Ali, M.A., Al-Hemaid, F.M., Lee, J., Hatamleh, A.A., Gyulai, G. and Rahman, M.O. 2015b. Unraveling systematic inventory of *Echinops* (Asteraceae) with special reference to nrDNA ITS sequence based molecular typing of *Echinops abuzinadianus*. Genet. Mol. Res. **14**(4): 11752–11762.

Ali, M.A., Gyulai, G. and Al-Hemaid, F. 2015c. Plant DNA Barcoding and Phylogenetics. LAP Lambert Academic Publishing, Germany.

Altschul, S.F., Gish, W., Miller, W., Myers, E.W. and Lipman, D.J. 1990. Basic local alignment search tool, J. Mol. Biol. **215**: 403–410.

Alvarez, I. and Wendel, J.F. 2003. Ribosomal ITS sequences and plant phylogenetic inference. Mol. Phyl. Evol. **29**: 417–434.

Baldwin, B.G., Sanderson, M.J., Porter, J.M., Wojciechowski, M.F., Campbell, C.S. and Donoghue, M.J. 1995. The ITS region of nuclear ribosomal DNA: a valuable source of evidence on angiosperm phylogeny. Ann. Miss. Bot. Gard. **82**: 247–277.

Chen, S., Yao, H., Han, J., Liu, C., Song, J., Shi, L., Zhu, Y., Ma, X., Gao, T., Pang, X., Luo, K., Li, Y., Li, X., Jia, X., Lin, Y. and Leon, C. 2010. Validation of the ITS2 region as a novel DNA barcode for identifying medicinal plant species. PLoS ONE **5**(1): e8613.

Devos, N., Barker, N.P., Nordenstam, B. and Mucina, L. 2010. A multi-locus phylogeny of *Euryops* (Asteraceae, Senecioneae) augments support for the "Cape to Cairo" hypothesis of floral migrations in Africa. Taxon **59**(1): 57–67.

Feliner, G.N., Larena, B.G. and Aguilar, J.F. 2004. Fine-scale geographical structure, intra-individual polymorphisms and recombination in nuclear ribosomal internal transcribed spacers in *Armeria* (Plumbaginaceae). Ann. Bot. **93**: 189–200.

Hall, T.A. 1999. BioEdit: a user-friendly biological sequence alignment editor and analysis program for Windows 95/98/ NT. Nucleic Acids Sym. Ser. **41**: 95–98.

Lee, J., Kim, S.Y., Park, S.H. and Ali, M.A. 2013. Molecular phylogenetic relationships among members of the family Phytolaccaceae *sensu lato* inferred from internal transcribed spacer sequences of nuclear ribosomal DNA. Genet. Mol. Res. **12**(4): 4515–4525.

Nordenstam, B. 1968. The genus *Euryops*. Part I. Taxonomy. Opera Bot. **20**: 1–409.

Nordenstam, B. 1969. Phytogeography of the genus *Euryops* (Compositae). A contribution to the phytogeography of southern Africa. Opera Bot. **23**: 1–77.

Nordenstam, B., Clark, V.R., Devos, N. and Barker, N.P. 2009. Two new species of *Euryops* (Asteraceae: Senecioneae) from the Sneeuberg, Eastern Cape Province, South Africa. S. African J. Bot.75: 144–152.

Pandey, A.K. and Ali, M.A. 2012. Intraspecific variation in *Panax assamicus* Ban. (Araliaceae) populations based on internal transcribed spacer (ITS1-5.8S-ITS2) sequences of nrDNA. Indian J. Biotech. **11**: 30–38.

Poczai, P., and Hyvönen, J. 2010. Nuclear ribosomal spacer regions in plant phylogenetics: problems and prospects. Mol. Biol. Rep. **37**(4): 1897–1912.

Prokopovich, C.D., Gregory, T.R. and Crease, T.J. 2003.The correlation between rDNA copy number and genome size in eukaryotes. Genome **46**: 48–50.

Szabo, Z., Gyulai, G., Humphreys, M., Horvath, L., Bittsánszky, A., Lagler, R. and Heszky, L. 2005. Genetic variation of melon (*C. melo*) compared to an extinct landrace from the Middle Ages (Hungary) I. rDNA, SSR and SNP analysis of 47 cultivars. Euphytica **146**: 87–94.

Swofford, D.L. 2002. PAUP: Phylogenetic Analysis Using Maximum Parsimony (and Other Methods). Version 4.0b 10. Sinauer, Sunderland, Massachusetts.

Tamura, K., Peterson, D., Peterson, N., Stecher, G., Nei, M. and Kumar, S. 2011. MEGA5: molecular evolutionary genetics analysis using maximum likelihood, evolutionary distance, and maximum parsimony methods. Mol. Biol. Evol. **28**(10): 2731–2739.

Thompson, J.D., Gibson, T.J., Plewniak, F., Jeanmougin, F. and Higgins, D.G. 1997. The Clustal X windows interface: flexible strategies for multiple sequence alignment aided by quality analysis tools. Nucleic Acids Res. **24**: 4876–4882.

White, T.J., Bruns, T., Lee, S. and Taylor, J. 1990. Amplification and direct sequencing of fungal ribosomal RNA genes for phylogenetics. *In*: Innis, M.A., Gelfand, D.H., Sninksky, J.J. and White, T.J. (Eds), PCR protocols: a guide to method and amplifications. Academic Press, San Diego, pp. 315–322.

Yao, H., Song, J., Liu, C., Luo, K., Han, J., Li, Y., Pang, X., Xu, H., Zhu, Y., Xiao, P. and Chen, S. 2010. Use of ITS2 region as the universal DNA barcode for plants and animals. PLoS ONE **5**(10): e13102.

5

STATUS OF *RESEDA PENTAGYNA* ABDALLAH & A.G. MILLER (RESEDACEAE) INFERRED FROM COMBINED NUCLEAR RIBOSOMAL AND CHLOROPLAST SEQUENCE DATA

M. AJMAL ALI[1], FAHAD M. AL-HEMAID, RITESH K. CHOUDHARY[2],
JOONGKU LEE[2], SOO-YONG KIM[2] AND M.A. RUB[3]

*Department of Botany and Microbiology, College of Science,
King Saud University, Riyadh 11451, Saudi Arabia*

Keywords: Reseda pentagyna; Resedaceae; Saudi Arabia; Endemic; ITS; *trn*L-F.

Abstract

The present study focuses on the status of *Reseda pentagyna* Abdallah & A.G. Miller (Resedaceae). The internal transcribed spacer (ITS) region of nuclear ribosomal DNA and chloroplast *trn*L-F gene of the questioned species were sequenced. The Basic Local Alignment Search Tool (BLAST) search showed maximum identity with *R. stenostachya*. The parsimony analysis of ITS, *trn*L-F and combined sequences data analyses revealed grouping of *Reseda* species consistent with established taxonomic sections of the genus, *R. pentagyna* showed proximity with *R. stenostachya* (100% bootstrap support), nested within the clade of section *Reseda*.

Introduction

The Resedaceae include six genera (i.e. *Caylusea* A. St.-Hil, *Ochradenus* Delile, *Oligomeris* Cambess., *Randonia* Coss., *Reseda* L. and *Sesamoides* Ortega) with approximately 85 species, and are widely distributed in the Old World, with a major center of species diversity in the Mediterranean basin (Martín-Bravo *et al.*, 2007). The members of the family Resedaceae has been traditionally considered closely related to Capparaceae and Brassicaceae; however, the Angiosperm Phylogeny Group placed it under the order Brassicales (APG III, 2009).

The genus *Reseda* consists of approximately 65 species, mostly restricted to the Mediterranean basin, while four of them (i.e. *Reseda alba* L., *R. lutea* L., *R. luteola* L. and *R. phyteuma* L.) are distributed throughout the world (Martín-Bravo *et al.*, 2007). The genus *Reseda* in Saudi Arabia is represented by seven species, *viz. R. alba, R. arabica* Boiss., *R. aucheri* Boiss., *R. lutea, R. muricata* C. Presl, *R. pentagyna* Abdallah & A.G. Miller and *R. sphenocleoides* Deflers (Chaudhary, 1999). Among these, *R. pentagyna* is endemic to Saudi Arabia, and reported to occur in Northern Hijaz mountain area, Wadi Sawawin and Tabuk of north western Saudi Arabia (Miller and Nyberg, 1994; Chaudhary, 1999; Llewellyn *et al.*, 2010). *R. stenostachya* is the most closely allied taxon to the endemic *R. pentagyna* which differs from the latter by presence of only 3-4 toothed capsules as compared to the 5-6 toothed capsules in the latter.

In the last two decades, the internal transcribed spacer sequences of nuclear ribosomal DNA has gained much attention, not only because of its efficacy in carrying out phylogeny of the plants at lower taxonomic level, but also to be considered as the most trusted markers available for the DNA barcoding of the plants. Even after facing criticism of its utility, this marker stands parallel

[1]Corresponding author. Email: majmalali@rediffmail.com
[2]International Biological Material Research Center, Korea Research Institute of Bioscience and Biotechnology, Daejeon-305 806, South Korea.
[3]National Herbarium & Genebank, National Agriculture & Animal Resources Research Center, Riyadh-11484, Saudi Arabia.

to the smartest genes available for the molecular phylogeny and plant DNA barcoding. Since the intrigued morphological similarities observed in between *R. pentagyna* and *R. stenostachya* (Miller and Nyberg, 1994; Chaudhary, 1999) we planned to carry out molecular phylogenetic analysis of internal transcribed spacer sequences (ITS) of nuclear ribosomal DNA and *trn*L-F sequences to confirm the species status of *R. pentagyna*.

Materials and Methods

The leaf material of *R. pentagyna* was collected from Wadi Sirr area of Saudi Arabia, and the taxonomic identification was confirmed through consultation of Flora of Saudi Arabia (Chaudhary, 1999) and protologue (Miller and Nyberg, 1994). Total genomic DNA was extracted using the DNeasy Plant Mini kit (QIAGEN, Valencia, CA, USA). The nuclear (internal transcribed spacer sequences of nuclear ribosomal DNA), and plastid (*trn*L-F) genes were amplified using AccuPower HF PCR PreMix (Bioneer, Daejeon, South Korea). The standard primers ITS (White *et al.*, 1990) and *trn*L-*trn*L-F (Taberlet *et al.*, 1991) were used for amplification and cycle sequencing. The amplified products were purified using PCR purification kit (SolGent, Daejeon, South Korea) prior to sequencing. The purified amplified products were sequenced using ABI PRISM 3730XL (Perkin-Elmer/Applied Biosystem, USA) following manufacturer's protocol. Each sample was sequenced in the sense and anti-sense direction. The nucleotide sequences of both the DNA strands (sense and anti-sense) were obtained and analyzed using Sequence Navigator (Perkin-Elmer/Applied Biosystems) to ensure accuracy of the base pair sequence.

For the molecular phylogenetic analysis, ITS and *trn*L-F sequences of a total of 36 related species of *Reseda* (comprising representative from all six sections i.e. *Glaucoreseda, Leucoreseda, Luteola, Neoreseda, Phyteuma* and *Reseda* as recognized by Martín-Bravo *et al.*, 2007) were retrieved from GenBank (Table 1). According to Martín-Bravo *et al.* (2007) *Oligomeris* arose within the ranks of *Reseda*; hence, sequences of *Oligomeris* were retrieved from GenBank, and were used as outgroup in the phylogenetic analyses (Table 1). Sequence alignments were performed using Clustal X, version 1.81 (Thompson *et al.*, 1997). Sequence alignments were subsequently adjusted manually using BioEdit (Hall, 1999). Gaps were treated as missing data in phylogenetic analyses. The voucher specimen (Chaudhary *et al.* 13704) of sequenced plant accession deposited at National Herbarium (RIY) of Saudi Arabia; and the generated sequences submitted in GenBank (Table 1). Maximum parsimony (MP) analysis was performed using PAUP* 4.0b10 (Swofford, 2002).

Results and Discussion

The combined length of ITS region (ITS1-5.8S-ITS2) in *Reseda pentagyna* was 634 bp. The ITS1 region was 261 bp (GC content 61%), the 5.8S gene was 162 bp (GC content 56%), and the ITS2 region was 211 bp (GC content 63%). The *trn*L-F sequence in *R. pentagyna* was 777 bp (GC content 33%). BLAST search of ITS sequence of *R. pentagyna* showed maximum identity (99%) with *R. stenostachya* followed by *R. aucheri* and *R. ellenbeckii* (95%), while *trn*L-F sequence showed maximum identity (100%) with *R. stenostachya* followed by *R. alphonsi, R. buhseana, R. gilgiana* and *R. sessilifolia* (97%). ITS sequence of *R. pentagyna* differs from *R. stenostachya* at position 67 and 75 in alignment, however, in *trn*L-F sequences, no base pair difference was observed in between sequence of *R. pentagyna* and *R. stenostachya*. Sequence characteristics and statistics of maximum parsimony trees derived from analyses of ITS, *trn*L-F and combined data are summarized in Table 2. The maximum parsimony tree derived from analysis of ITS and *trn*L-F sequence revealed comparatively week bootstrap support than combined analysis; and therefore, only the maximum parsimony trees topology derived from analysis of combined sequence data is discussed here.

Table 1. Plant accessions used for the molecular phylogenetic analysis of *Reseda pentagyna*.

Taxa	GenBank Accession No.	
	ITS	*trn*L-F
Ingroup		
Sect. *Glaucoreseda*		
1. *Reseda battandieri* Pit.	DQ987183	DQ987045
2. *R. complicata* Bory	DQ987172	DQ987046
3. *R. glauca* L.	DQ987182	DQ987040
4. *R. gredensis* (Cutanda & Willk.) Müll.-Arg.	DQ987174	DQ987047
5. *R. virgata* Boiss. & Reut.	DQ987177	DQ987048
Sect. *Luteola*		
6. *R. luteola* L.	DQ987187	DQ987050
Sect. *Leucoreseda* Subsect. *Leucoreseda*		
7. *R. alba* L.	DQ987198	DQ987053
8. *R. attenuata* Ball	DQ987201	DQ987057
9. *R. gayana* Boiss.	DQ987205	DQ987055
10. *R. undata* L.	DQ987203	DQ987056
11. *R.valentina* Pau	DQ987207	DQ987059
Sect. *Leucoreseda* Subsect. *Erythroreseda*		
12. *R. suffruticosa* Loefl.	DQ987210	DQ987062
Sect. *Neoreseda*		
13. *R. ellenbeckii* Perkins	DQ987110	DQ986998
14. *R. telephiifolia* (Chiov.) Abdallah & de Wit	DQ987128	DQ986994
Sect. *Phyteuma*		
15. *R. alopecuros* Boiss.	DQ987139	DQ987028
16. *R. arabica* Boiss.	DQ987132	DQ987029
17. *R. collina* Müll.-Arg.	DQ987136	DQ987031
18. *R. diffusa* Ball	DQ987141	DQ987033
19. *R. inodora* Rchb.	DQ987142	DQ987030
20. *R. odorata* L.	DQ987133	DQ987026
21. *R. orientalis* (Müll.-Arg.) Boiss.	DQ987137	DQ987025
22. *R. phyteuma* L.	DQ987146	DQ987032
Sect. *Reseda*		
23. *R. alphonsi* Müll.-Arg.	DQ987108	DQ987005
24. *R. amblycarpa* Fresen.	DQ987125	DQ987001
25. *R. aucheri* Boiss.	DQ987123	DQ986989
26. *R. buhseana* Müll.-Arg.	DQ987119	DQ987004
27. *R. crystallina* Webb & Berthel.	DQ987088	DQ987021
28. *R. gilgiana* Perkins	DQ987114	DQ986999
29. *R. lanceolata* Lag.	DQ987099	DQ987015
30. *R. lutea* L.	DQ987094	DQ987018
31. *R. pentagyna* Abdallah & Miller	JX867260	JX867261
32. *R. sessilifolia* Thulin	DQ987127	DQ986995
33. *R. sphenocleoides* Deflers	DQ987117	DQ986993
34. *R. stenostachya* Boiss.	DQ987156	DQ987007
35. *R. stricta* Pers.	DQ987103	DQ987013
36. *R. urnigera* Webb	DQ987098	DQ987014
37. *R. viridis* Balf. f.	DQ987130	DQ986996
Outgroup		
38. *Oligomeris dipetala* (Aiton) Turcz.	DQ987168	DQ987037
39. *O. dregeana* (Müll. Arg.) Müll.-Arg.	DQ987166	DQ987038
40. *O. linifolia* (Vahl) J.F. Macbr.	DQ987165	DQ987039

Table 2. Summary of sequence characteristics and MP trees derived from analyses of ITS, *trn*L-F and combined data.

Characters	ITS	*trn*L-F	Combined data
Number of taxa included in analysis (including outgroup)	40	40	40
Sequence characteristics			
Length of sequenced	627-639	698-785	1325-1424
Aligned length	644	955	1622
Parsimony informative	92	146	433
Tree characteristics			
Number of trees	334	323	1299
Length	339	327	1305
CI (Consistency Index)	0.643	0.832	0.656
RI (Retention Index)	0.885	0.926	0.860
RC (Rescaled Consistency Index)	0.569	0.770	0.564
HI (Homoplasy Index)	0.478	0.318	0.446

The bootstrap strict consensus tree resulted from combined sequence data analysis has been shown in Fig. 1. The study revealed the grouping of *Reseda* species according to previously recognized taxonomic sections, which is consistent with earlier report (Martín-Bravo *et al.*, 2007). Moreover, *R. pentagyna* nested within the clade of the section *Reseda*, and showed proximity (100% bootstrap support) with morphologically similar *R. stenostachya*. The ITS sequence of *R. pentagyna* (which was described based on 5-6 toothed capsule characters) differs from morphologically allied *R. stenostachya* (3-4 toothed capsule) at aligned position 67 (C in *R. pentagyna* but missing nucleotide in *R. stenostachya*) and 75 (C in *R. pentagyna*, T in *R. stenostachya*) possibly due to nucleotide polymorphism, a known features of ITS sequences of nrDNA.

Bentham and Hooker (1862) reported *Reseda* as a polymorphic genus with not more than 30 existing species. Latter, Abdallah and de Wit (1978) updated the list with some addition, and emphasized the need of experimental taxonomical research to get a strong support for the delimitation of species. Muller (1864) also described the variations in the morphology of leaf blades of *Reseda* that might be arranged in various manners and could be entire, crenate to ternately or pinnately (or rarely bi-pinnately) lobed. The occurrence of brachycarpous or macrocarpous capsules in *Reseda* is a known feature (Muller, 1864). Under various ecological conditions, plants may show certain morphological changes, *viz. R. lutea* shows change in the proximity of the veins in the lamina (Abdallah and de Wit, 1978). Further, the emergence of indumentums depends more or less on the moisture content present in the plant. In dry condition, these hairs can shrink, flatten or curl; while in wet conditions, they appear as blisters, or a scabrid, or muricated surface. As variations in fruit size within the same species usually do occur, therefore it cannot be taken as a strong taxonomic character for species level delimitation (Donald, 1988); and thus, the wide degree of variation in quantitative fruit-spine characters limits their use taxonomically. The proximity of questioned sequenced material with *R. stenostachya* in the MPTs indicates the quantitative differences of tooth characters or the variable trait which limits its use in species delimitation; therefore, we herein propose the merger of *R. pentagyna* into *R. stenostachya*.

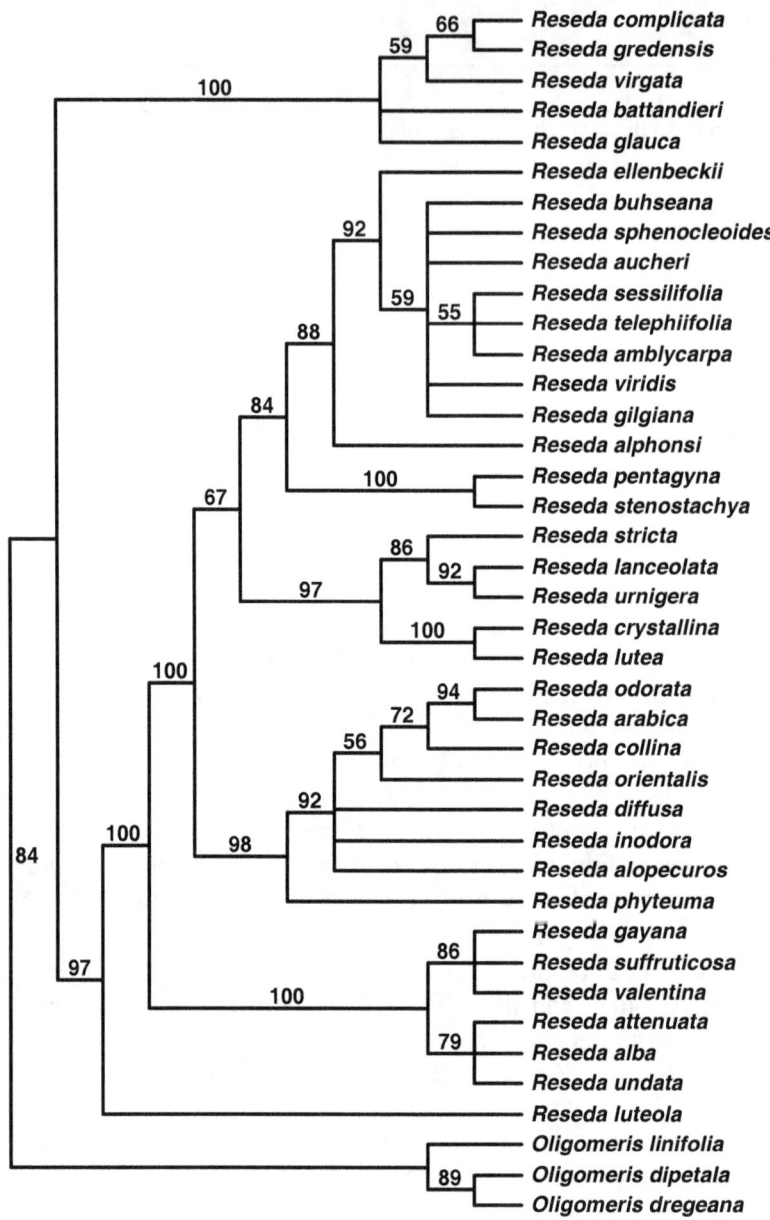

Fig. 1. Bootstrap strict consensus tree inferred from combined sequence data analysis of internal transcribed spacer (ITS) sequence of nuclear ribosomal DNA and *trn*L-F region. The Bootstrap strict consensus tree of 1299 maximally parsimonious trees (MPTs) with a total length of 1305 steps, a consistency index (CI) of 0.656, a homoplasy index (HI) of 0.446, rescaled consistency index (RC) of 0.564 and a retention index (RI) of 0.860. Bootstrap values greater than 50% in 1000 bootstrap replicates are shown above lines.

Acknowledgement

The authors would like to extend their sincere appreciation to the Deanship of Scientific Research at King Saud University for its funding of this research through the Research Group Project No. RGP-VPP-195.

References

Abdallah, M.S. and de Wit, H.C.D. 1978. The Resedaceae: a taxonomical revision of the family. Meded. Landbouwhoogeschool, Wageningen, p. 78.

APG III 2009. An update of the angiosperm phylogeny group classification for the orders and families of flowering plants. Bot. J. Linn. Soc. **161**: 105-121.

Bentham, G., and Hooker, J.D. 1862. Genera Plantarum. Reeve, Williams & Norgate, London.

Chaudhary, S. 1999. Resedaceae. *In:* Chaudhary, S. (Ed.), Flora of the Kingdom of Saudi Arabia. Ministry of Agriculture and Water, National Herbarium, National Agriculture and Water Research Center, Riyadh, Saudi Arabia, pp. 536-543.

Donald, H.L. 1988. The evolution of achene morphology in *Ceratophyllum* (Ceratophyllaceae), II. Fruit variation and systematic of the "spiny-margined" group. Syst. Bot. **13**(1): 73-86.

Felsenstein, J. 1985. Confidence limits on phylogenies: an approach using the bootstrap. Evolution. **39**: 783-791.

Hall, T.A. 1999. BioEdit: A user-friendly biological sequence alignment editor and analysis program for Windows 95/98/NT. Nuc. Acids Symp. Ser. **41**: 95-98.

Llewellyn, O.A., Hall, M., Miller, A.G., Al-Abbasi, T.M., Al-Wetaid, A.H., Al-Harbi, R.J., Al-Shammari, K.F. and Al-Farhan, A. 2010. Important plant areas in the Arabian Peninsula: 1. Jabal Qaraqir. Edinb. J. Bot. **67**: 37-56.

Martín-Bravo, S., Meimberg, H., Luceño, M., Märkl, W., Valcárcel, V., Bräuchler, C., Vargas, P. and Heubl, G. 2007. Molecular systematics and biogeography of Resedaceae based on ITS and *trn*L-F sequences. Mol. Phylogenet. Evol. **44**: 1105-1120.

Miller, A.G. and Nyberg, J.A. 1994. Studies in the flora of Arabia: XXVII some new taxa from the Arabian Peninsula. Edinb. J. Bot. **51**(1): 33-47.

Müller, A.J. 1864. Resedaceae. *In:* De Candolle, A.P. (Ed.) Prodromus Systematis Naturalis Regni Vegetabilis, Victor Masson, Paris **16**(2): 548-589.

Swofford, D.L. 2002. PAUP* (v. 4.0b10). Phylogenetic analysis using parsimony (* and other methods). Sinauer Associates, Sunderland.

Taberlet, P., Gielly, L., Pautou, G. and Bouvet, J. 1991. Universal primers for amplification of three non-coding regions of chloroplast DNA. Plant Mol. Biol. **17**: 1105-1109.

Thompson, J.D., Gibson, T.J., Plewniak, F., Jeanmougin, F. and Higgins, D.G. 1997. The Clustal_X windows interface: Flexible strategies for multiple sequence alignment aided by quality analysis tools. Nucleic Acids Res. **24**: 4876-4882.

White, T.J., Bruns, T., Lee, S. and Taylor, J. 1990. Amplification and direct sequencing of fungal ribosomal RNA genes for phylogenetics. *In:* Innis, M.A., Gelfand, D.H., Sninksky, J.J. and White, T.J. (Eds) PCR protocols: a guide to method and amplifications. Academic Press, San Diego, California, pp. 315-322.

6

MOLECULAR EVOLUTIONARY RELATIONSHIPS OF *EUPHORBIA SCORDIFOLIA* JACQ. WITHIN THE GENUS INFERRED FROM ANALYSIS OF INTERNAL TRANSCRIBED SPACER SEQUENCES

FAHAD M.A. AL-HEMAID, M. AJMAL ALI[1], JOONGKU LEE[2], SOO-YONG KIM[3]
AND M. OLIUR RAHMAN[4]

*Department of Botany and Microbiology, College of Science, King Saud University,
Riyadh 11451, Saudi Arabia*

Keywords: *Euphorbia scordifolia*; Euphorbiaceae; ITS; Genotyping.

Abstract

The present study explored molecular phylogenetic analysis of 28 species of *Euphorbia* L. for the identification and establishment of molecular evolutionary relationships of *Euphorbia scordifolia* Jacq. within the genus based on the internal transcribed spacers (ITS) sequences (ITS1-5.8S-ITS2) of nuclear ribosomal DNA (nrDNA). The sequence similarity search using Basic Local Alignment Search Tool (BLAST) of the ITS sequence of *E. scordifolia* showed the closest sequence similarity to *E. supina* Raf. The analysis of ITS sequence data revealed four major clades consistent with subgeneric classifications of the genus. Molecular data support placement of *E. scordifolia* in the subgenus *Chamaesyce*.

Introduction

The genus *Euphorbia* L. (Euphorbiaceae) comprising ca. 2000 species, which is one of the largest genera of the flowering plants (Frodin, 2004; Riina *et al.*, 2013). The main molecular phylogenetic studies of *Euphorbia* species have addressed the overall phylogeny of the genus, with its four subgeneric clades of *Rhizanthium*, *Esula*, *Euphorbia*, and *Chamaesyce* (Steinmann and Porter, 2002; Bruyns *et al.*, 2006; Park and Jansen, 2007; Zimmermann *et al.*, 2010). In Saudi Arabia, the genus *Euphorbia* is represented by 38 species. Of them, *E. scordifolia* Jacq. is distributed in Cape Verde Island, Ethiopia, Somalia, Sudan, Yemen and also in western region of Saudi Arabia (Abedin *et al.*, 2001). The morphological characters of *E. scordifolia* overlap with *E. supina* Raf. (Abedin *et al.*, 2001).

The internal transcribed spacers (ITS) sequence of nuclear ribosomal DNA region including the 5.8S gene is the most widely used molecular marker to infer phylogenetic relationships among plant species (Baldwin *et al.*, 1995; Ali *et al.*, 2014). Although reliance on nrDNA ITS sequence as the sole source of phylogenetic evidence has come under criticism because of certain features of its evolution; however, it remains the most efficient locus for generating species-specific phylogenetic inferences and genotyping in most groups of plants (Ali *et al.*, 2013, 2014, 2015). While searching for DNA sequences of *E. scordifolia* in GenBank as a part of a research for genotyping of unresolved taxonomic status of flowering plants of Saudi Arabia, it was found that *E. scordifolia* have not previously been sequenced. A perusal of taxonomic literature revealed that

---

[1]Corresponding author. Email: majmalali@rediffmail.com
[2]Department of Environment and Forest Resources, Chungnam National University, 99 Daehak-ro, Yuseong-gu, Daejeon 34134, South Korea.
[3]International Biological Material Research Center, Korea Research Institute of Bioscience and Biotechnology, Daejeon 305 806, South Korea.
[4]Department of Botany, University of Dhaka, Dhaka 1000, Bangladesh.

the molecular evolutionary relationships of *E. scordifolia* distributed in Saudi Arabia is also unknown. Therefore, the present study aims at molecular genotyping of *E. scordifolia* based on ITS sequence of nrDNA.

Materials and Methods

Taxon sampling:

Leaf materials of *E. scordifolia* were collected from the herbarium specimens [voucher- Al-Rawshan, altitude 1122 m, 19.08.1978, Don Bermant 146] housed at National Herbaium & Genebank, National Agriculture & Animal Resources Research Center, Ministry of Agriculture, Riyadh, Saudi Arabia (RIY); and the taxonomic identification was confirmed through consultation of Flora of Saudi Arabia (Abedin *et al.*, 2001).

DNA extraction, amplification and sequencing:

Total genomic DNA was extracted using Qiagen DNeasy Plant Mini Kit (Valencia, CA, USA). ITS sequences of nuclear ribosomal DNA were amplified using AccuPower HF PCR PreMix (Bioneer, Daejeon, South Korea) and primer ITS1 (5$'$-GTCCACTGAACCTTATCATTT AG-3$'$) and ITS4 (5$'$-TCCTCCGCTTATTGATATGC-3$'$) of White *et al.* (1990) via polymerase chain reaction (PCR). Each 20 µl volumes of PCR premix contained 2 µl of 10x buffer, 300 µM dNTPs, 1 µl of a 10 pM solution of each primer and 1 unit of HF DNA polymerase. One round of amplification consisted of denaturation at 94 °C for 5 min, followed by 40 cycles of denaturation at 94 °C for 1 min, annealing at 49 °C for 1 min and extension at 72 °C for 1 min, and a final extension for 5 min at 72 °C. PCR products were purified with the SolGent PCR Purification Kit-Ultra (SolGent, Daejeon, South Korea) prior to sequencing. The sequencing reaction was performed in a 10 µl final volume with the BigDye Terminator cycle sequencing kit (Perkin-Elmer, Applied Biosystems). Cycling conditions included an initial denaturation at 94 °C for 5 min, followed by 30 cycles of 96 °C for 10 s, 50 °C for 5 s, and 60 °C for 4 min. The sequenced products were precipitated with 17 µl of deionized sterile water, 3 µl of 3 M NaOAc, and 70 µl of 95% EtOH. The capillary gel electrophoresis was conducted with Long Ranger Single Packs (FMC BioProducts) by an ABI 3100 automated DNA sequencer (Perkin-Elmer, Applied Biosystems). The sequences were analyzed by ABI Sequence Navigator (Perkin-Elmer/Applied Biosystems). Nucleotide sequences of both DNA strands were analyzed to ensure accuracy. The sequences were subjected to BLAST-searched (Altschul *et al.*, 1990) by NCBI server (http://blast.ncbi.nlm.nih.gov/Blast.cgi).

Phylogenetic analysis:

ITS sequences of nrDNA of 28 species of *Euphorbia* (Table 1) were retrieved from GenBank database of National Center for Biotechnology Information (www.ncbi.nlm.nih.gov). *Neoguillauminia cleopatra* and *Dichostemma glaucescens* were chosen as outgroup taxa according to previous work (Barres *et al.*, 2011) and were retrieved from GenBank (Table 1). Sequence alignment was performed using CLUSTAL X version 1.81 (Thompson *et al.*, 1997). Sequence alignment was subsequently adjusted manually using BioEdit (Hall, 1999). Gaps were treated as missing data in phylogenetic analyses. The generated sequences were submitted to GenBank (Table 1). The boundaries between the ITS1, 5.8S and ITS2 gene for *E. scordifolia* were determined in the aligned data matrix, and were exported as a Nexus file and subsequently analysed using Maximum Parsimony (MP) and Maximum Likelihood (ML) methods by MEGA5 (Tamura *et al.*, 2011). The distribution and pattern of nucleotide substitution in all sequences was investigated using HYPERMUT (Rose and Korber, 2000).

Table 1. Plant accessions used for the molecular phylogenetic analysis of *Euphorbia scordifolia.*

Group	Subgenus	Taxon	GenBank Accession No.
Ingroup	*Rhizanthium*	*Euphorbia antso* Denis	AF537579
		Euphorbia atrispina N.E. Br.	AF537568
		Euphorbia balsamifera Ait.	AF537571
		Euphorbia clava Jacq.	AF537569
		Euphorbia namuskluftensis L.C. Leach	AF537562
		Euphorbia obesa Hook. f.	AF537566
	Esula	*Euphorbia aphylla* Brouss.	AF537540
		Euphorbia dendroides L.	AF537539
		Euphorbia peplus L.	AF537532
		Euphorbia schimperi C. Presl	AF537537
		Euphorbia schimperiana Hochst. *ex* A. Rich.	JN207816
	Euphorbia	*Euphorbia abdelkuri* Balf. f.	AF537458
		Euphorbia beharensis Leandri	AJ508983
		Euphorbia cylindrifolia Marn.-Lap. & Rauh	AJ508955
		Euphorbia drupifera Thonn.	AF537480
		Euphorbia epiphylloides Kurz	AF537484
		Euphorbia milii Des Moul.	AJ508974
		Euphorbia ramipressa Croizat	AF537481
		Euphorbia supina Raf.	EU659773
		Euphorbia teke Schweinf. *ex* Pax	AF537485
	Chamaesyce	*Euphorbia fulgens* Karw. *ex* Klotzsch	AF537404
		Euphorbia graminea Jacq.	AF537410
		Euphorbia heterophylla L.	GU214931
		Euphorbia ipecacuanhae L.	AF537397
		Euphorbia leucocephala Lotsy	GU214932
		Euphorbia misera Benth.	AF537383
		Euphorbia pulcherrima Willd. *ex* Klotzsch	GU214943
		Euphorbia scordifolia Jacq.	KR704890
		Euphorbia sphaerorhiza Benth.	AF537412
Outgroup		*Neoguillauminia cleopatra* (Baill.) Croizat	AF537581
		Dichostemma glaucescens Pierre	AF537584

Results and Discussion

The combined length of ITS region (ITS1-5.8S-ITS2) in *E. scordifolia* was 642 bp. The ITS1 region was 266 bp (GC content 53%), the 5.8S gene was 162 bp long (GC content 56%), and the ITS2 region was 213 bp (GC content 58%). The BLAST search of ITS sequence of *E. scordifolia* showed high identity level (95%) with *E. humifusa* Willd. followed by *E. glyptosperma* Engelm., *E. maculata* L., *E. tettensis* Klotzsch and *E. meganaesos* Featherm. Parsimony analysis of the entire ITS region resulted in five maximally parsimonious trees, the consistency index was 0.491, the retention index was 0.709, and the composite index was 0.367 (0.348) for all sites and parsimony-informative sites (in parentheses). There were a total of 499 positions in the final dataset, of which 223 were parsimony informative. The phylogenetic tree recovered by the analyses provided a clear resolution of taxon included in the analysis at the subgeneric level. *Eupphorbia scordifolia* nested within the clade of the subgenus *Chamaesyce*. The ML analyses recovered tree topology similar to MPT; and therefore, only the ML topology is presented here (Fig. 1). A total of 36 specific nucleotide differences, i.e. 19 in ITS1 and 17 in ITS2 region were detected between *E. scordifolia* and *E. supina* (Table 2).

Table 2. Differences of DNA base pairs between the ITS sequences of *Euphorbia supina* and *E. scordifolia*.

Specific nucleotide differences					
Position in sequence alignment	ITS1		Position in sequence alignment	ITS2	
	E. supina	*E. scordifolia*		*E. supina*	*E. scordifolia*
18	G	A	3	T	C
41	T	T	22	C	T
45	C	G	25	T	C
56	G	T	37	-	G
93	C	T	49	C	T
106	T	-	56	A	R
113	C	T	74	T	C
135	A	C	94	T	C
136	A	T	126	T	C
137	A	T	146	A	G
147	T	C	151	C	A
148	G	T	163	C	T
149	C	T	170	T	A
208	C	T	173	G	A
212	C	T	174	A	T
215	C	T	191	T	C
232	T	C	192	G	A
254	G	A			
258	G	A			

The Tandem Repeats Finder (Benson, 1999) was used to detect repeats in the ITS sequences. Differences in substitution rates can discriminate functional forms of pseudogenes (Buckler and Holtsford, 1996a,b). The analysis using the program HYPERMUT showed excessive levels of G =>A mutations which indicates that all differences arose from a single substitution sequence. The result was compared to the reference sequences and their physical locations along the sequences were graphically illustrated (Fig. 2).

The use of DNA sequences to identify organisms has been proposed as a more efficient approach than traditional and morphological taxonomic parameters (Tautz *et al.*, 2003). In fact, the recent development in DNA molecular systematic techniques including molecular hybridization, cloning, restriction endonuclease digestions and DNA sequencing and phylogenetic theory have changed the epitome of species identification as well as our understanding of the relationships among organisms at various levels in the tree of life which has been advanced greatly

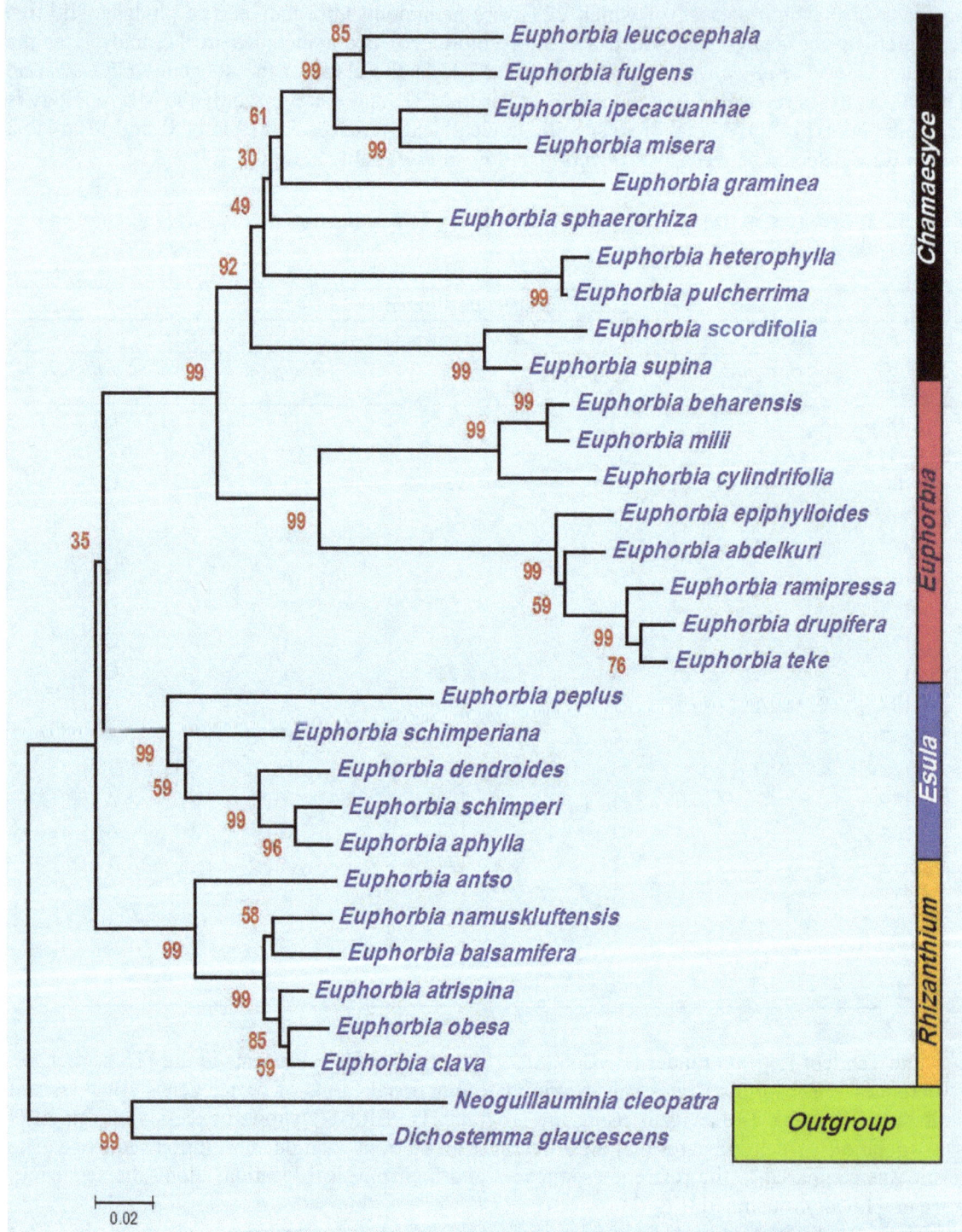

Fig.1. A maximum likelihood (ML) tree inferred from analysis of sequence data of internal transcribed spacer (ITS) region of nuclear ribosomal DNA. Bootstrap values (1000 × replicates) are indicated.

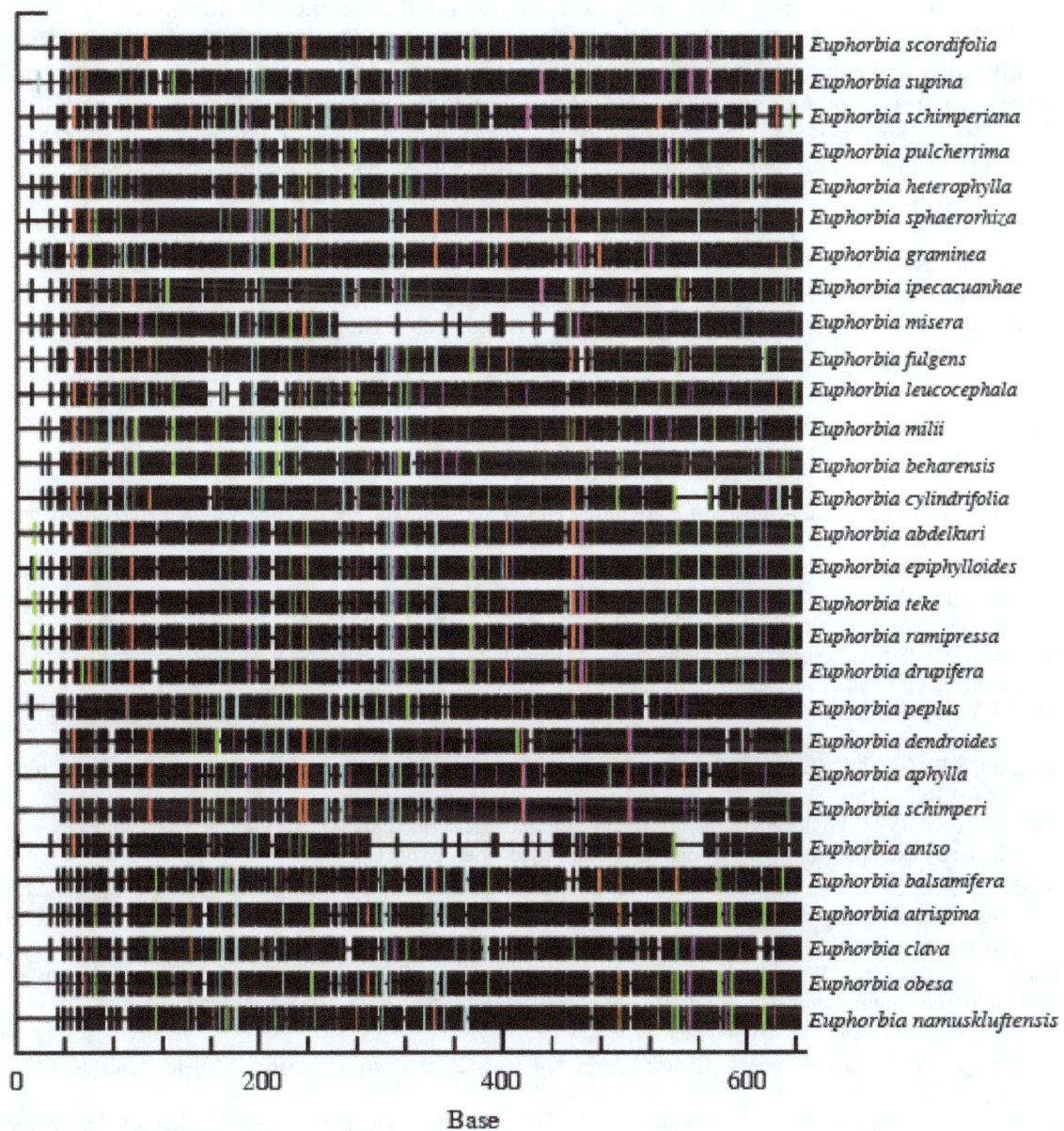

Fig. 2. Schematic illustration of the distribution of substitution sites across the ITS region obtained from 29 species of *Euphorbia*, using *Dichostemma glaucescens* as reference (red = GG > AG, cyan = GA > AA, green = GC > AC, magenta = GT > AT, black = not G > A transition, yellow = gap).

(Ali *et al.*, 2014). From the first report of the utility of the nrDNA ITS sequence in plants (Baldwin, 1992), it has been extensively used to distinguish even very closely related species (Chen *et al.*, 2010; Yao *et al.*, 2010). Moreover, during the last two decades, the nrDNA ITS sequence has gained much attention as smartest gene available for the molecular signature of a taxon (Ali *et al.*, 2013).

The present study is the first report of inferring the nrDNA ITS based molecular genotyping of the *E. scordifolia*. Since, the majority of the species of the genus *Euphorbia* have to be sequenced; the present study will nevertheless help in DNA barcoding / molecular identification of *E.*

scordifolia as well as it will also participate in addressing the complete phylogeny of the genus *Euphorbia*. The DNA barcodes show promise in providing a practical, standardized, species-level identification tool that can be used for biodiversity assessment, life history, ecological studies and forensic analysis (Szabó *et al.*, 2005; Mansour *et al.*, 2009; Gyulai *et al.*, 2012; Ali *et al.*, 2014, 2015). Hence, the nrDNA ITS sequence of *E. scordifolia* will be of immense importance in barcoding of the genus *Euphorbia* in particular, and in the analysis of plant biodiversity of Saudi Arabia in general.

Acknowledgement

The authors would like to extend their sincere appreciation to the Deanship of Scientific Research at King Saud University for funding this research through the Research Group Project No. RGP-VPP-195.

References

Abedin, S., Mossa, J.S., Al-Said, M.S. and Al-Yahya, M.A. 2001. Euphorbiaceae. *In*: Chaudhary, S. (Ed.), Flora of Saudi Arabia, Ministry of Agriculture and Water, National Herbarium, National Agriculture and Water Research Center, Riyadh, Saudi Arabia, **II**(1): 291–395.

Ali, M.A., Al-Hemaid, F.M.A., Choudhary, R.K., Lee, J., Kim, S.Y. and Rub, M.A. 2013. Status of *Reseda pentagyna* Abdallah & A.G. Miller (Resedaceae) inferred from analysis of combined nuclear ribosomal and chloroplast sequence data. Bangladesh J. Plant Taxon. **20**(2): 233–238.

Ali, M.A., Gyulai, G., Norbert, H., Balázs, K., Al-Hemaid, F.M.A., Pandey, A.K. and Lee, J. 2014. The changing epitome of species identification - DNA barcoding. Saudi J. Biol. Sci. **21**(3): 204–231.

Ali, M.A., Gyulai, G. and Al-Hemaid, F. 2015. Plant DNA Barcoding and Phylogenetics. LAP Lambert Academic Publishing, Germany, pp. 1–298.

Altschul, S.F., Gish, W., Miller, W., Myers, E.W. and Lipman, D.J. 1990. Basic local alignment search tool. J. Mol. Biol. **215**: 403–410.

Baldwin, B.G. 1992. Phylogenetic utility of the internal transcribed spacers of nuclear ribosomal DNA in plants: an example from the Compositae. Mol. Phylogenet. Evol. **1**: 3–16.

Baldwin, B.G., Sanderson, M.J., Porter, J.M., Wojciechowski, M.F., Campbell, C.S. and Donoghue, M.J. 1995. The ITS region of nuclear ribosomal DNA: a valuable source of evidence on angiosperm phylogeny. Ann. Miss. Bot. Gard. **82**: 247–277.

Barres, L., Vilatersana, R., Molero, J., Susanna, A. and Galbany-Casals, M. 2011. Molecular phylogeny of *Euphorbia* subg. *Esula* sect. *Aphyllis* (Euphorbiaceae) inferred from nrDNA and cpDNA markers with biogeographic insights. Taxon **60**(3): 705–720.

Benson, G. 1999. Tardem repeats finder: a program to analyze DNA sequence. Nucleic Acids Res. **27**: 573–580.

Bruyns, P.V., Mapaya, R.J. and Hedderson, T. 2006. A new subgeneric classification of *Euphorbia* (Euphorbiaceae) based on ITS and *psb*A–*trn*H sequence data. Taxon **55**: 397–420.

Buckler, E.S. and Holtsford, T.P. 1996a. *Zea* ribosomal repeat evolution and mutation patterns. Mol. Biol. Evol. **13**: 623–632.

Buckler, E.S. and Holtsford, T.P. 1996b. *Zea* systematics: ribosomal ITS evidence. Mol. Biol. Evol. **13**: 612–622.

Chen, S., Yao, H., Han, J., Liu, C., Song, J., Shi, L., Zhu, Y., Ma, X., Gao, T., Pang, X., Luo, K., Li, Y., Li, X., Jia, X., Lin, Y. and Leon, C. 2010. Validation of the ITS2 region as a novel DNA barcode for identifying medicinal plant species. PLoS ONE **5**(1): e8613.

Frodin, D. 2004. History and concepts of big plant genera. Taxon **53**: 753–776.

Gyulai, G., Horváth, L., Lágler, R. and Holly, L. 2012. The Hungarian gene bank collections of common millet (*Panicum miliaceum*) and the application for conservation genetics. European J. Plant Sci. Biotech. **6**(SI2): 69–102.

Hall, T.A. 1999. BioEdit: a user-friendly biological sequence alignment editor and analysis program for Windows 95/98/ NT. Nuc. Acids. Sym. Ser. **41**: 95–98.

Mansour, A., Ismail, H.M., Ramadan, M.F. and Gyulai, G. 2009. Comparative genotypic and phenotypic analysis of tomato (*Lycopersicon esculentum*) cultivars grown under two different seasons in Egypt. African J. Plant Sci. Biotechnol. **3**(1): 73–79.

Park, K.-R. and Jansen, R.K. 2007. A phylogeny of Euphorbieae subtribe Euphorbiinae (Euphorbiaceae) based on molecular data. J. Plant Biol. **50**: 644–649.

Riina, R., Peirson, J.A., Geltman, D.V., Molero, J., Frajman, B., Pahlevani, A., Barres, L., Morawetz, J.J., Salmaki, Y., Zarre, S., Kryukov, A., Bruyns, P.V. and Berry, P.E. 2013. A worldwide molecular phylogeny and classification of the leafy spurges, *Euphorbia* subgenus *Esula* (Euphorbiaceae). Taxon **62**(2): 316–342.

Rose, P.P. and Korber, B.T. 2000. Detecting hypermutations in viral sequences with an emphasis on G A hypermutation. Bioinformatics **16**: 400–401.

Szabó, Z., Gyulai, G., Humphreys, M., Horváth, L., Bittsánszky, A., Lágler, R. and Heszky, L. 2005. Genetic variation of melon (*C. melo*) compared to an extinct landrace from the middle ages (Hungary) I. rDNA, SSR and SNP analysis of 47 cultivars. Euphytica **146**: 87–94.

Steinmann, V.W. and Porter, J.M. 2002. Phylogenetic relationships in Euphorbieae (Euphorbiaceae) based on ITS and *ndh*F sequence data. Ann. Miss. Bot. Gard. **89**: 453–490.

Tamura, K., Peterson, D., Peterson, N., Stecher, G., Nei, M. and Kumar, S. 2011. MEGA5: molecular evolutionary genetics analysis using maximum likelihood, evolutionary distance, and maximum parsimony methods. Mol. Biol. Evol. **28**(10): 2731–2739.

Tautz, D., Arctander, P., Minelli, A., Thomas, R.H. and Vogler, A.P. 2003. A plea for DNA taxonomy. Trends Ecol. Evol. **18**(2): 70–74.

Thompson, J.D., Gibson, T.J., Plewniak, F., Jeanmougin, F. and Higgins, D.G. 1997. The Clustal X windows interface: flexible strategies for multiple sequence alignment aided by quality analysis tools. Nucleic Acids Res. **24**: 4876–4882.

White, T.J., Bruns, T., Lee, S. and Taylor, J. 1990. Amplification and direct sequencing of fungal ribosomal RNA genes for phylogenetics. *In*: Innis, M.A., Gelfand, D.H., Sninksky, J.J. and White, T.J. (Eds), PCR Protocols: A Guide to Method and Amplifications. Academic Press, San Diego, pp. 315–322.

Yao, H., Song, J., Liu, C., Luo, K., Han, J., Li, Y., Pang, X., Xu, H., Zhu, Y., Xiao, P. and Chen, S. 2010. Use of ITS2 region as the universal DNA barcode for plants and animals. PLoS ONE **5**(10): e13102.

Zimmermann, N.F.A., Ritz, C.M. and Hellwig, F.H. 2010. Further support for the phylogenetic relationships within *Euphorbia* L. (Euphorbiaceae) from nrITS and *trn*L–*trn*F IGS sequence data. Plant Syst. Evol. **286**: 39–58.

7

MOLECULAR SYSTEMATICS OF SOME BIFURCATE HAIRY SECTIONS IN *ASTRAGALUS* L. (FABACEAE) AS INFERRED FROM NUCLEAR AND CHLOROPLAST DNA SEQUENCES

REZA SHEIKHAKBARI-MEHR[1], ALI ASGHAR MAASSOUMI[2] AND
SHAHROKH KAZEMPOUR OSALOO[3]

Department of Biology, Faculty of Science, University of Qom, Qom, Iran

Keywords: *Astragalus*; cpDNA; Fabaceae; nrDNA ITS; Phylogeny.

Abstract

In this study, 38 species belonging to some bifurcate hairy sections of *Astragalus* L. were analyzed phylogenetically, using nuclear and plastid DNA sequences. Based on our results, *Astragalus* sect. *Dissitiflori* DC. with the inclusion of the members of section *Erioceras* Bunge, formed a monophyletic group. The members of sect. *Ornithopodium* Bunge and *Onobrychoidei* DC. were located together within a highly supported monophyletic clade, apart from other sections studied, on the basis of the present molecular data.The positioning of the enigmatic, recently established species, *A. juladakensis* Maassoumi, within the sect. *Dissitiflori* was verified. In addition, our results showed that *A. pravitzii* Podl., which had been already transferred to sect. *Ornithopodium*, belongs to the section *Dissitiflori*.

Introduction

Astragalus L. (family Fabaceae, subfamily Faboideae) is among the largest genera of the flowering plants containing up to 3000 species of herbs and small shrubs (Maassoumi, 2005; Lewis *et al.,* 2005).The south-western and central Asia are considered as the main centers of biodiversity for the Old World *Astragalus* (Lock and Simpson, 1991). Infrageneric and sectional classification of *Astragalus* was first carried out by De Candolle (1825) with the description of 14 sections, a number then increased by Boissier (1843).However, the first comprehensive classification of the Old World *Astragalus* was presented by Bunge (1868), with the description of 150 sections in 10 subgenera. The current distinction of 150 and 93 sections belonging to the Old World and New World *Astragalus* respectively indicates that *Astragalus* is a complex genus within Angiosperms (Barneby, 1964; Podlech, 1986). These sections are distinguished based on some morphological characters such as stem features, stipules connation, leaf shape, inflorescence and legume features (Maassoumi, 2000). There are more than 800 species of *Astragalus* in Iran, which has a high endemism rate of 65% (Podlech, 1999; Maassoumi, 2005).

Astragalus sect. *Dissitiflori* DC is one of the largest sections among bifurcate hairy *Astragalus*, with more than 150 species in the world (Ranjbar, 2004) and about 20 species in Iran (Podlech *et al.,* 2010). Ghahreman *et al.,* (1996) transferred *A. viridis* Bunge and *A. dendroproselius* Rech. f. from *Dissitiflori* to the section *Cystodes* Bunge. Later on, these two species along with *A. aestimabilis* Podlech were moved to sect. *Corethrum* Bunge (Maassoumi, 2005). According to Maassoumi (2005), sect. *Corethrum* Bunge is closely related to sect. *Dissitiflori* but differs from that especially in having oblong elliptic pods and long spreading hairs

[1] Corresponding author. Email: r.sheikhakbari@qom.ac.ir; reza.sheikhakbari@gmail.com
[2] Botany Division, Research Institute of Forests and Rangelands, Tehran, Iran.
[3] Department of Botany, Faculty of Biological Sciences, Tarbiat Modares University, Tehran, Iran.

on fruit. Therefore, this section was recorded for Iran by transferring three aforementioned species from sect. *Dissitiflori* based on their fruit characteristics (Maassoumi, 2005).

Astragalus sect. *Erioceras* Bunge is closely related to the *Dissitiflori* and has been probably evolved by shortening of stem in the latter (Ranjbar and Karamian, 2002). The species of sect. *Erioceras* are xerophytes and more or less caespitose in contrast to many other bifurcate hairy sections.

Sect. *Cytisodes* Bunge which was originally established by Bunge (1868) with one species is now presented by 17 species (Podlech, 2010). This section was included in Flora of Iran after discovery of a new species, *A. gigantirostratus* Maassoumi *et al.,* (1999). Later on, Podlech (2004) published another new species belonging to sect. *Cytisodes* in Iran. Recently Maassoumi (2005) transferred *A. zoshkensis* Ghahremani, from section *Dissitiflori* to the *Cytisodes*. However, according to the latest revision of *Astragalus* in Flora Iranica, section *Cytisodes* has only two species in Iran (Podlech *et al.,* 2010).

The only inclusive molecular phylogenetic analyses of the Old World *Astragalus*, using nrDNA ITS and in part plastid gene *ndh*F sequences are those of Kazempour Osaloo *et al.,* (2003, 2005). Based on these studies, large sections of *Astragalus* such as *Incani* DC., *Cenanthrum* Bunge and *Ammodendron* Bunge formed monophyletic groups. In contrast, sections *Chlorostachys* Bunge, *Hystrix* Bunge, *Heterodonthus* Bunge, *Hymenostegis* Bunge, *Acidodes* Bunge, *Rhacophorus* Bunge and Iranian endemic section *Leucocercis* Bunge are not monophyletic. Moreover, monophyly of sections *Dissitiflori* DC., *Erioceras* Bunge, *Laguropsis* Bunge, *Macrocystis* Popov, *Stenonychium* Bunge, and *Onobrychoidei* DC. remained unresolved (Kazempour Osaloo *et al.,* 2005).

The aims of this study were: 1) to evaluate the phylogenetic status of sections *Dissitiflori* and *Erioceras* in Iran, on the basis of nrDNA and cpDNA sequences, and 2) to find the correct position of some problematic species i.e. *A. juladakensis* Maassoumi (2007), *A. pravitzii* Podlech (2001), and *A. zoshkensis* Ghahremani-nejad (2003)) related to these sections.

Materials and Methods

Taxon sampling

A total of 38 taxa were chosen as in-group for nrDNA ITS, and cpDNA *trn*H-*psb*A, *mat*K (as partial), and *trn*T-*trn*Y sequence analyses (Table 1). The in-group mainly belonged to sections *Dissitiflori* and *Erioceras*. In order to determine the situation of some controversial species, a number of representatives pertaining to the closely related sections such as *Ornithopodium* Bunge, *Onobrychoidei*, and *Cytisodes* were introduced in the analyses. *Astragalus stocksii* Bunge and *A. frigidus* (L.) A. Gray was chosen as outgroups following previous molecular phylogenetic studies in the Old World *Astragalus* (Kazempour Osaloo *et al.,* 2003, 2005; Sheikh Akbari-Mehr *et al.,* 2012a, 2012b). The cpDNA sequences for majority of in-group and ITS for 16 species (marked with an asterisk at Table 1) are published here for the first time.

DNA extraction, PCR and Sequencing

Total genomic DNA was extracted from dry leaves of individual plants, deposited in Central Herbarium of Iran (TARI) and Ferdowsi University of Mashhad Herbarium (FUMH), following the modified CTAB procedure of Doyle and Doyle (1987). The complete nrDNAITS+5.8S region was amplified using primers ITS4 of White *et al.,* (1990) and ITS5m of Sang *et al.,* (1997). The cpDNA *mat*K (partial), *trn*H-*psb*A and *trn*T-*trn*Y regions were amplified using primers trnK-F and matK-R (Wojciechowski *et al.,* 2004), trnH and psbA (Tate and Simpson, 2003) and trnT and

Table 1. Taxa included in the molecular analyses and their voucher specimens. Sequences obtained from GenBank marked with an asterisk.

Species	Voucher no.	GenBank accession no.			
		ITS	*trn*T/Y	*trn*H/*psb*A	*mat*K
Astragalus argyroides Beck.	Mozaffarian & Freitag, 28538(TARI)	*AB721936	LC129368	LC129321	*AB727543
A. aucheri Boiss.	Mottaghi, 1061(TARI)	*AB721937	-	LC129319	-
A. argentocalyx Ali *ex* Podl.	Ghahremaninejad & Joharchi, 34738(TARI)	LC129287	-	LC129323	LC129310
A. eburneus Born. & Gauba	Mozaffarian, 44936(TARI)	*AB721938	LC129353	LC129318	LC129299
A. husseinovii Rezazade	Maassoumi & Safavi, 8721(TARI)	*AB721939	-	LC129341	LC129308
A. juratzkanus Freyn & Sint.	Maassoumi & Pakravan, 72351(TARI)	*AB721940	LC129366	LC129347	LC129306
A. melanocalyx Boiss. & Buhse	Noruzi & Feizi, 5860(TARI)	*AB721941	LC129357	LC129335	LC129298
A. baraftabensis Maass.& Podl.	Tayebi, 4458(TARI)	*AB721942	LC129352	LC129317	LC129307
A. nigrolineatus Sirj. & Rech.f.	Faghihnia & Zangooee,29042(FMUH)	*AB721943	LC129367	LC129324	LC129297
A. pravitzii Podl.	Foroughi,2183(TARI)	*AB721944	LC129358	LC129332	*AB727544
A. ruscifolius Boiss.	Mozaffarian & Freitag, 28640(TARI)	*AB721945	LC129369	LC129320	*AB727545
A. sitiens Bge.	Wendelbo & Foroughi, 11270(TARI)	*AB721947	LC129362	LC129333	LC129305
A. saadatabadensis Podl.	Grant, 15784(TARI)	*AB721946	-	LC129330	LC129292
A. sumbari Popov	Wendelbo & Foroughi, 11063(TARI)	*AB721948	LC129370	LC129316	-
A. xiphidium Bge.	Youssefi, 7611(TARI)	*AB721949	-	LC129336	LC129296
A. juladakensis Maassoumi	Maassoumi, 39383 (TARI)	*AB721950	-	LC129340	LC129295
A. aestimabilis Podl.	Dehshiri, 38523(TARI)	*AB721951	-	-	-
A. dendroproselius Rech.f.	Dehshiri, 30231(TARI)	*AB721952	-	LC129322	LC129293
A. viridis Bunge.	Moussavi, 1152(TARI)	*AB721953	-	LC129345	-
A. zoshkensis F. Ghahremani	Mozaffarian, 77059(TARI)	*AB721954	LC129360	LC129331	LC129294
A. gigantirostratus Maassoumi *et al.,*	Maassoumi & al., 72339(TARI)	*AB721955	-	LC129338	-
A. anacamptus Bunge.	Emadzadeh & al., 35908(FUMH)	* AB721956	LC129365	LC129327	LC129311
A. djenarensis Sirj. & Rech.f.	Joharchi & Zangooee, 1100(TARI)	*AB721957	LC129355	LC129342	LC129303
A. stocksii Bunge.	Foroughi, 10802(TARI)	*AB051966	*AB741437	-	*AB741345
A. frigidus(L.) A. Gray	5732(TARI)	*AM943381	*AB741412	-	*AB741320

Table 1 contd.

Species	Voucher no.	GenBank accession no.			
		ITS	*trn*T/Y	*trn*H/*psb*A	*mat*K
A. bifoliolatus Sirj. & Rech.f.	Asadi & Amirabadi, 9342(TARI)	LC129283	LC129361	-	LC129309
A. alamliensis Rech.f.	Asadi, 84461(TARI)	LC129284	-	LC129334	-
A. catacamptus Bunge	Dini & bazargan, 5328(TARI)	LC129288	-	LC129329	LC129312
A. keredjensis Podl.	Asadi, 82404(TARI)	LC129291	LC129355	LC129328	-
A. neosytinii Ranjbar	Asadi, 84571(TARI)	LC129280	LC129354	LC129343	LC129301
A. nubicola Podl.	Wendelbo, 11165(TARI)	LC129289	-	LC129339	-
A. pakravaniae Podlech & Maassoumi	Asadi & Maassoumi, 55534(TARI)	LC129286	-	LC129337	-
A. pentanthus Boiss.	Maroofi, 1917(TARI)	LC129290	LC129363	LC129325	LC129302
A. sympiliecarpus Rech.f.	Asadi & Maassoumi, 83362(TARI)	LC129285	LC129351	LC129344	LC129300
A. versipilus Rech. f. & Koie	Asadi & Amirabadi, 84615(TARI)	LC129281	LC129356	LC129346	LC129313
A. brachyodontus Boiss.	Asadi & Wendelbo, 27666(TARI)	*AB727530	-	-	*AB727537
A. jodostachys Boiss. & Buhse	Abuhamzeh & Maassoumi, 45496(TARI)	*AB727532	-	-	*AB727539
A. gotkschaicus Grossh.	Asadi & Foroughi, 13756(TARI)	*AB727515	LC129372	LC129350	LC129315
A. teheranicus Boiss. & Hohen.	Babakhanlou & Amin, 15069(TARI)	*AB727523	LC129371	LC129349	LC129314
A. ahangarensis Zarre & Podl.	Abbasi & Amirabadi, 4416(TARI)	LC129282	LC129359	LC129326	LC129304

trnY (Demesure *et al.,* 1995), respectively. The total volume of amplification reaction was 25 μl, made up of 18 μl deionized water, 2.5 μl of 10× PCR buffer, 2.5 μl of 2.5 mM dNTPs, 0.5 μl of each primer (5 pmol μl-1), 0.25 μl (5 units per μl) of *Taq*DNA polymerase and0.75 μl of template DNA. The PCR profile for ITS consisted of 2.5 min at 95°C for pre-denaturation followed by 27cycles of 1 min at 95°C for denaturation, 45 sec at53.7°C for primer annealing and 50 sec at 72°C for primer extension, and a final primer extension of 7 min at 72°C.PCR procedure for amplification of three cpDNA regions was as follows: 3 min at 94°C, 35 cycles of 1 min at 94°C, 1 min at 51–64°C, 1.5 min at72°C, and terminal elongation of 7 min at 72°C.PCR products were directly used for sequencing reactions. Sequencing of the nrDNA ITS and cpDNA fragments were performed using an ABI 3130Genetic DNA Analyzer (Applied Biosystems, USA).

Sequence alignment

Sequences of nuclear and plastid DNA were edited by BioEdit package version7 (Hall 1999). The sequence alignment was carried out using ClustalX (Larkin *et al.,* 2007) and adjusted manually. Indel positions were treated as missing data.

Phylogenetic analyses

Maximum parsimony

Sequenced nuclear and plastid fragments were analyzed separately and in combination, using maximum parsimony method (MP) as implemented in the PAUP* version 4.0b10 (Swofford, 2002). Multiple tree searches were conducted using heuristic search options that included random addition sequences (100 replicates), holding five trees per replicate, and tree bisection-reconnection (TBR) branch swapping with retention of multiple parsimonious trees (Maxtrees = 25000). Bootstrap (BP) support values (Felsenstein, 1985) were calculated using a full heuristic search with 1000 replicates, each with a simple addition sequence and TBR branch swapping. Uninformative characters were excluded from analyses. Parsimony trees were not shown here.

Bayesian analyses

All datasets separately and in combination, were analyzed using Bayesian inference (BI) as implemented in MrBayes version 3.1.2 (Ronquist and Huelsenbeck, 2003). The incongruent length difference (ILD) test was performed to evaluate the combinability of the all DNA regions studied (Farris *et al.,* 1995). Appropriate evolutionary models for analyzing sequences were selected using the MrModeltest2 (Nylander, 2004) based on the Akaike information criterion (AIC) (Posada and Buckley 2004). K80+I+G, GTR+I+G, GTR+I, and F81+G were chosen as the models that best fit the datasets of nrDNA ITS, *trn*H-*psb*A, *mat*K and *trn*T-*trn*Y respectively. In combined dataset, various sequences were included as separate partitions. BI analyses were run for two million generations, using Markov chain Monte Carlo search. MrBayes performed two simultaneous analyses starting from different random trees (N runs=2) each with four Markov chains and trees sampled at every 100 generations. In all analyses average standard deviation of split frequencies had dropped significantly below 0.01 after completion of the generations. Once reaching the stationary phase, trees were collected and after burning in one fourth of them, used to build a 50% majority rule consensus tree accompanied with posterior probability (PP) values. Trees were showed using TreeGraph2 (Stöver and Müller, 2010).

Results and Discussion

nrDNA ITS dataset analyses

The average length of aligned nrDNA ITS fragment was 596. Three nucleotide sites, of which 60 sites were parsimony informative. The Bayesian tree with posterior probabilities (PP) and bootstrap values is similar to that of MP analysis (Fig. 1). Based on these analyses, four species belonging to the sections *Ornithopodium* and *Onobrychoidei* were located at the base of tree as a sister group to a large assemblage of five subclades. *Astragalus juladakensis* was placed at the base of this group. Members of sections *Dissitiflori* and *Erioceras* plus *Cytisodes* were well intermixed and formed several subclades within a large monophyletic group (Fig. 1). Although relationships among these subclades were not resolved, each one is supported with moderately to highly bootstrap or PP values.

cpDNA and combined datasets analyses

Parsimony trees obtained from three single cpDNA and the combined cpDNA plus ITS datasets, were topologically identical to those of Bayesian analyses. The length and composition of DNA sequences as well as the tree statistics from the single and combined analyses have been summarized in Table 2. In *trn*H-*psb*A tree, *A. tehranicus* Boiss. & Hohen. and *A. goktschaicus* Grossh. belonging to the sect. *Onobrychoidei* were united in a highly supported subclade (PP= 1) and placed at base of the tree as a sister to the remaining species (Fig. 2). Again, the members of

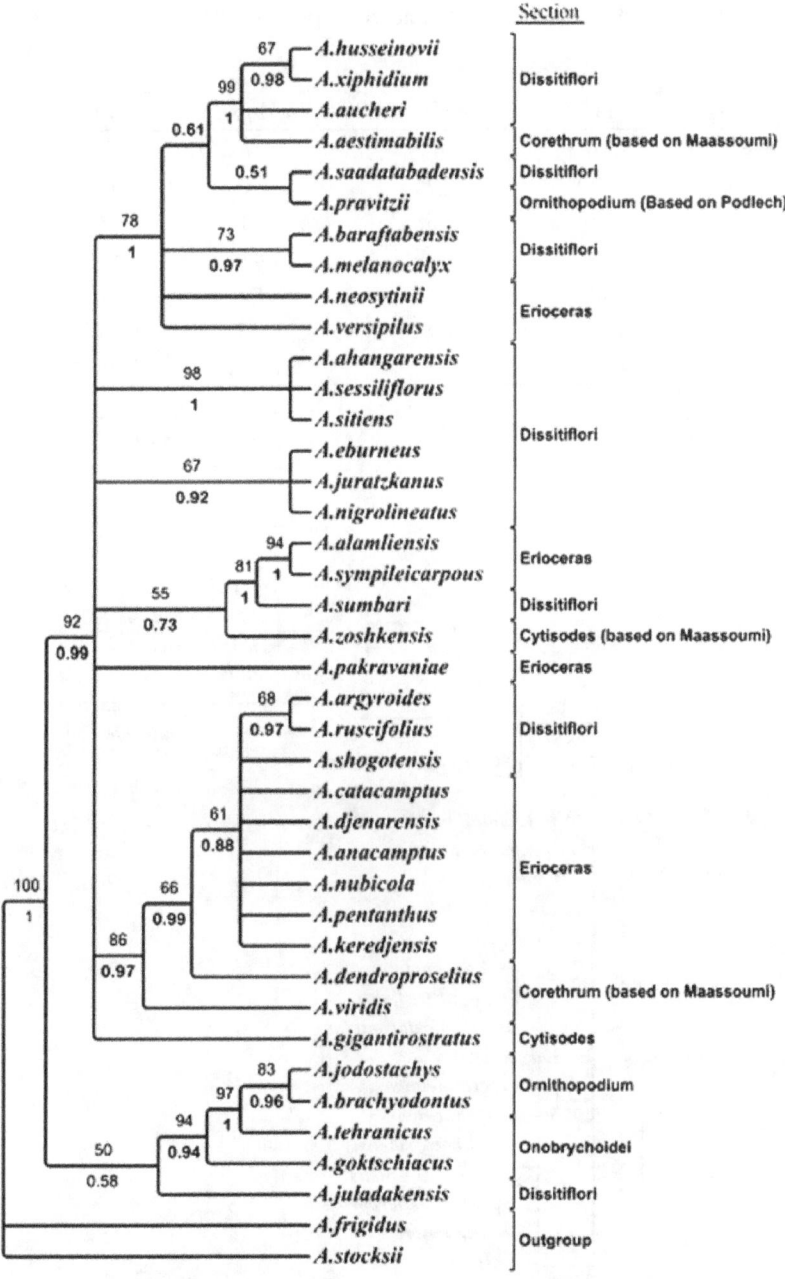

Fig.1. Fifty percent majority rule consensus tree resulting from Bayesian analysis of the nrDNA ITS dataset. Numbers above and below branches are bootstrap values and posterior probabilities, respectively.

sections *Dissitiflori* and *Erioceras* plus some controversial species (i.e. *A. zoshkensis*, *A. aestimabilis* Podl., *A. dendroproselius* Rech. f. and *A. viridis* Bunge) were intermixed within a large polytomic assemblage (Fig. 2). In the *matK* tree, species sampled from two sections *Onobrychoidei* and *Ornithopodium* revealed a highly supported group (BS= 80%, PP= 0.95) and placed as a sister to the members of other sections. The remaining species, in this tree as well as two other cpDNA trees, placed together within a polytomic large clade (Fig. 3). *trnT-trnY* region was not amplified in some of in-groups due to difficulties with the PCR. However, the topology of the tree obtained from this sequence was similar to the other trees in general (tree not shown here).

Table 2. Dataset and tree statistics from separate and combined analyses of the nuclear and three chloroplast regions.

Data sets	ITS	*trn*T/*trn*Y	*trn*H/*psb*A	*mat*K	combined
Nucleotide sites (average)	596.3	629	397.7	931	2554
Variable sites	120	76	82	61	337
Informative characters	60	58	44	18	178
Number of MPTs	10	39	6494	68	398
Length of MPTs	86	74	80	29	335
CI of MPTs	0.756	0.824	0.637	0.828	0.670
RI of MPTs	0.882	0.911	0.839	0.891	0.719

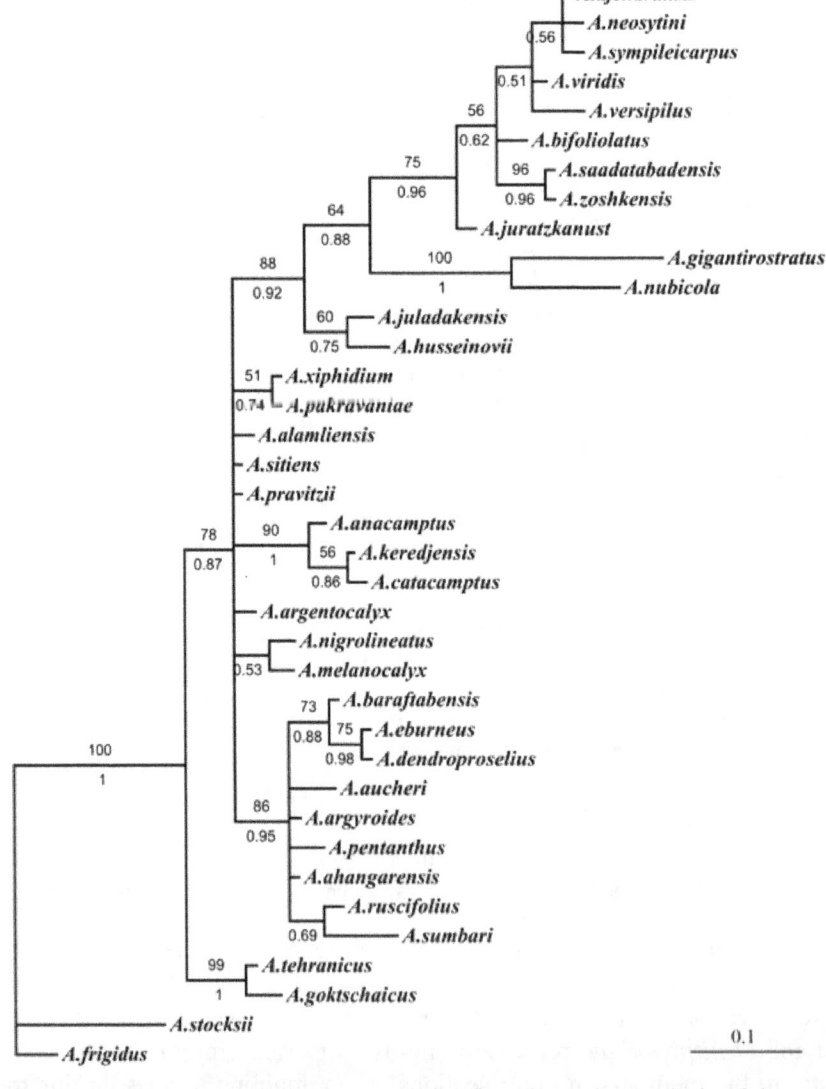

Fig. 2. Fifty percent majority rule consensus tree resulting from Bayesian analysis of the *trn*H/*psb*A dataset. Numbers above and below branches are bootstrap values and posterior probabilities, respectively.

ILD test suggested that the four datasets were slightly incongruent (*P*=0.01). Following the suggestions of several authors that the ILD test may be unreliable (Seelanan *et al.*, 1997; Wiens, 1998; Yoder *et al.*, 2001), we decided to combine these datasets. The DNA fragments which had not been sequenced for some species in this study were treated as missing data in the combined dataset. The topology of the resulted tree (Fig. 4) was roughly the same as those of single dataset trees, with the exception that resolution, bootstrap and PP values were higher. The combined tree

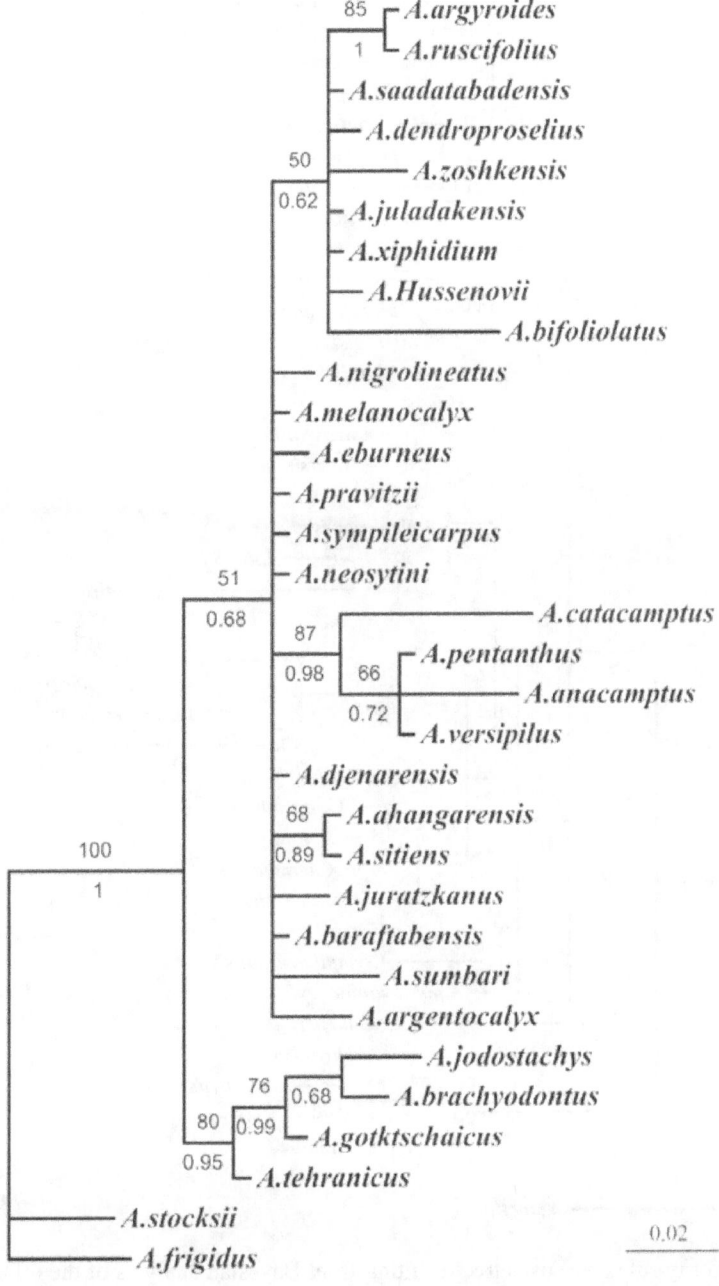

Fig. 3. Fifty percent majority rule consensus tree resulting from Bayesian analysis of the cpDNA *mat*K dataset. Numbers above and below branches are bootstrap values and posterior probabilities, respectively.

was composed of two obvious clades among in-groups studied. At base of the tree, four species belonging to the sections *Onobrychoidei* and *Ornithopodium* were separated from other in groups and formed a highly supported clade as a sister group to the remaining species (Fig. 4). The next main clade was composed of two clades, each of successive subclades including the members of sections *Dissitiflori* and *Erioceras* and their closely related taxa. The relationships of these subclades were well resolved (Fig. 4).

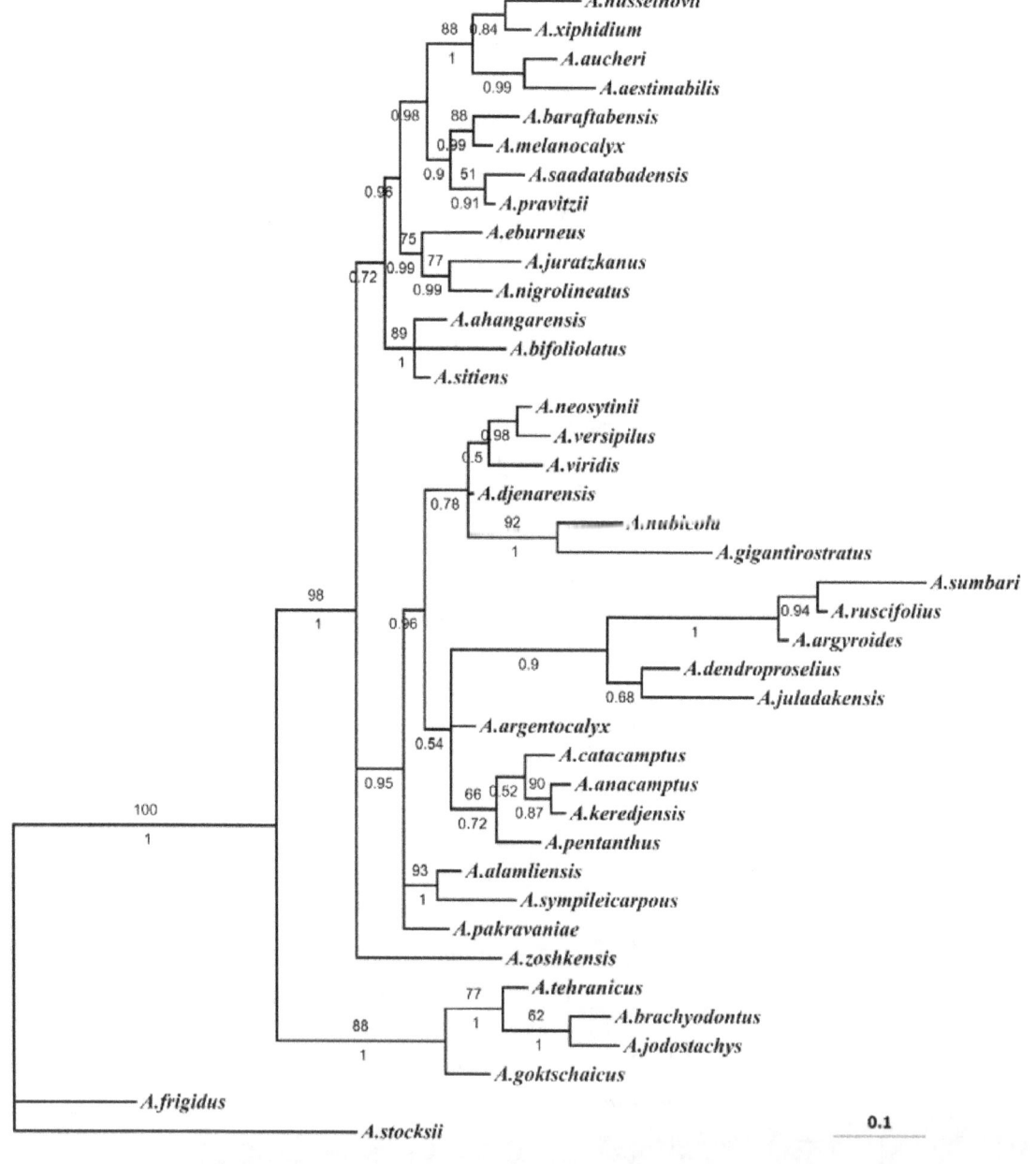

Fig. 4. Fifty percent majority rule consensus tree resulting from Bayesian analysis of the nrDNA and cpDNA combined dataset. Numbers above and below branches are bootstrap values and posterior probabilities, respectively.

Among different datasets analyzed here, relationships of species were well resolved on the ITS and combined trees. *Astragalus* sect. *Dissitiflori* is one of the largest sections of the genus including more than 40 species in the Iranian Plateau (Podlech *et al.*, 2010). Among bifurcate hairy *Astragalus*, the members of *Dissitiflori* are distinguished by some features including stem with long internodes, linear pod and asymmetrical and gibbous calyx at the base (Ghahremani-nejad, 2004; Sheikh Akbari *et al.*, 2012a). It seems that this section belongs to a group of medifixed hairy *Astragalus* including *A.* sect. *Cystodes*, *A.* sect. *Erioceras*, *A.* sect. *Cystium* Bunge, *A.* sect. *Cremoceras* Bunge and *A.* sect. *Trachycercis* Bunge (Ranjbar 2004). This idea is also supported partially with molecular evidences (Kazempour Osaloo *et al.*, 2005; Sheikh Akbari *et al.*, 2012b). Molecular phylogenetic analyses of the present study showed that the members of sections *Erioceras* and *Cytisodes* in Iran, were intermixed with those of section *Dissitiflori* and located within a large assemblage (Fig. 4).

A. juladakensis, which was recently introduced as a new species belonging to the section *Dissitiflori* (Maassoumi, 2007), revealed some affinity to the members of *Onobrychoidei* based on ITS sequences and nested at the base of ITS tree, as a sister to the remaining species (Fig. 1). Based on our previous phylogenetic study on the sect. *Dissitiflori* (based on ITS), this species revealed a separated position among other members of the section and its affinity to the sect. *Dissitiflori* remained questionable (Sheikh Akbari Mehr *et al.*, 2012b). Despite these results, *A. juladakensis*, was placed beside the other members of sect. *Dissitiflori* on the basis of our cpDNA and combined datasets analyses (Fig. 4). On the other hand, this species along with *A. husseiovii* Rezazade was united within a moderately supported subclade within sect. *Dissitiflori*, based upon morphological features (Sheikh Akbari *et al.*, 2012a); hence, the positioning of this species within the section *Dissitiflori* is verified.

A. pravitzii Podl. and *A. saadatabadensis* Podl. formed a sister subclade within section *Dissitiflori*, on the basis of ITS and combined trees. After introducing *A. pravitzii* as a new species from sect. *Dissitiflori* (Podlech, 2001), Podlech and Sytin (2010) moved it to the sect. *Ornithopodium*. In accordance with previous morphological data analysis (Sheikh Akbari Mehr *et al.*, 2012a), our present molecular data revealed that this taxon is a member of sect. *Dissitiflori* (Figs 1, 4).

According to Gontscharov *et al.* (1946) and Maassoumi (2005), A. sect. *Corethrum* is closely related to the sect. *Dissitiflori* but differs with that in having asymmetrical long hairs on calyx and pod shape. Three species (*A. aestimabilis*, *A. dendroproselius* and *A. viridis*) belonging to the sect. *Dissitiflori* were separated from the section and introduced as the members of newly recorded section *Corethrum* for Iran, based on having ovate-elliptic pods and asymmetrical standing indumentum on calyx (Maassoumi, 2005). However, in accordance with Podlech and Zarre (2013), our present molecular dataset analyses revealed that these taxa belong to the sect. *Dissitiflori*. sect. *Erioceras* is characterized by a short stem, prostrate habit, asymmetrical long hairs, oblong elliptic pods and rupturing of calyx (Maassoumi, 2005). It seems that sect. *Erioceras* has been evolved by reducing of stem length in sect. *Dissitiflori* (Ranjbar and Karamian, 2002). However, our results obtained from single and combined molecular datasets revealed no distinction between two sections. The members of sect. *Erioceras* have adapted to arid and windy sub-mountainous regions. They are distributed in arid central and north-eastern of Iran. The evolution of prostrate habit and dense and long hairs within section *Erioceras* is likely an adaptive behaviour due to its environmental conditions.

Section *Cytisodes* is a small section among bifurcate hairy *Astragalus* and is distinguished by their short stem internodes, calyx with standing hairs and long beak on the pod (Bunge, 1868). Maassoumi *et al.* (1999) introduced a new species from eastern part of Elburz Mountains, showing the features of sect. *Cytisodes*, and named *A. gigantirostratus*. Occurrence of this species in the

Hyrcanian province astonished the authors, because known species of the section are all confined to the Turkestanian floristic province of the Irano-Turanian region. Later on, Podlech (1999) introduced *A. neyshaburensis* Podl. as a new species from sect. *Cytisodes* in Iran. Maassoumi (2005) moved *A. zoshkensis* from section *Dissitiflori* to the *Cytisodes* based on calyx hairs and pod features. However, in agreement with a recent morphological study (Sheikh Akbari Mehr *et al.*, 2012a), our present molecular results revealed that these species are placed within section *Dissitiflori* and it is recommended that section *Cytisodes* is best to be retreated after complementary studies.

In summary, different genomic sequences revealed that the sect. *Dissitiflori* with the inclusion of the members of section *Erioceras* as well as members of *Cytisodes* in Iran, formed a monophyletic group. The present results indicated that taxa which had been transferred from sect. *Dissitiflori* have to be returned to the section, and from this point of view, sect. *Corethrum* has no representative in Iran and this result is in accordance with Podlech *et al.*, (2010) and Podlech and Zarre (2013) classifications. Our findings showed that delimitation of sect. *Dissitiflori* needs to be revised. Indeed, beside the increase of samples, the analysis of type specimen of aforementioned sections seems to be necessary to assess exact taxonomic situation of taxa discussed above.

Acknowledgments

We are grateful to the directors and curators of the Central Herbarium of Iran (TARI) and Herbarium of Ferdowsi University (FUMH) for the loan of materials and collections.

References

Barneby, R. 1964. Atlas of North American *Astragalus*. Memoirs of the New York Botanical Garden **13**: 1-1188.

Boissier, E. 1843. Diagnoses Plantarum Orientalium Novarum, Ser. I, Part. 2. Typographia Ferd Ramboz, Genevae, pp. 1-115.

Bunge, A. 1868. Generis Astragali Species Gerontogeae. Académie Impériale des Sciences, St. Pétersburg, pp. 1-254.

De Candolle, A.P. 1825. Prodromus Systematis Naturalis, Regni Vegetabilis, Vol **2**. *Astragaleae*. Argentorati et Londii, Parisiis, pp. 1-644.

Demesure, B., Sodzi, N. and Petit, R.J. 1995. A set of universal primers for amplification of polymorphic non-coding regions of mitochondrial and chloroplast DNA in plants. Molecular Ecology **4**: 129-131.

Doyle, J.J. and Doyle ,J. L. 1987. A rapid DNA isolation procedure for small quantities of fresh leaf tissue. Phytochemical Bulletin **19**: 11-15.

Ekici, M., Akan, H. and Aytac, Z. 2011.Taxonomic revision of *Astragalus* section Onobrychoidei DC. (Fabaceae) in Turkey. Turkish Journal of Botany **35**: 1-73.

Farris, J.S., Kallersjo, M., Kluge, A.G. and Bult, C. 1995. Testing significance of incongruence. Cladistics **10**: 315-319.

Felsenstein, J. 1985. Confidence limits on phylogenies: an approach using the bootstrap. Evolution **39**: 783-791.

Ghahreman, A., Pakravan, M. and Maassoumi, A. A. 1996. Note on the genus *Astragalus* (sect. *Xiphidium*) in Iran. Iranian Journal Botany 7(1): 45-50.

Ghahremani-nejad, F. 2003. *Astragalus zoshkensis* (Fabaceae), a new species from Iran. Annales Botanici Fennici **40**: 117-121.

Ghahremani-nejad, F. 2004. The sections of *Astragalus* L. with bifurcating hairs in Iran. Turkish Journal of Botany **28**: 101-117.

Gontscharov, N.F., Borissova, A.G., Gorschkova, S.G., Popov, M.G. andVassilczenko, I.T. 1946. *Astragalus* L. *In*: Komarov, V.L. and Shishkin,B.K. (Eds), Flora USSR. Vol. **12**. Editio Academiae Scientiarum, Leningrad, Russia, pp. 1-681.

Hall, T.A. 1999. BioEdit: a user-friendly biological sequence alignment editor and analysis program for Windows 95/98/NT. Nucleic Acids Symposium Series **41**: 95-98.

Kazempour Osaloo, S., Maassoumi, A.A. and Murakami, N. 2003. Molecular systematics of the genus *Astragalus* L. (Fabaceae): phylogenetic analyses of nuclear ribosomal DNA internal transcribed spacers and chloroplast gene *ndh*F sequences. Plant Systematics and Evolution **242**: 1-32.

Kazempour Osaloo, S., Maassoumi, A.A. and Murakami, N. 2005. Molecular systematics of the Old World *Astragalus* (Fabaceae) as inferred from nrDNA ITS sequence data. Brittonia **57**: 367-381.

Larkin, M.A., Blackshields, G., Brown, N.P., Chenna, R., McGettigan, P.A., McWilliam, H., Valentin, F., Wallace ,I.M., Wilm, A., Lopez, R., Thompson, J.D., Gibson, T.J. and Higgins, D.G.2007. Clustal Wand Clustal X version 2.0. Bioinformatics **23**:2947-2948.

Lewis, G.P., Schrire, B.D., Mackinder, B.A. and Lock, M. 2005. Legumes of the world. Kew Publishing, London, UK, pp. 577.

Lock, J.M. and Simpson, K. 1991. Legumes of West Asia: A Check-List. Kew Publishing.

Maassoumi, A. A., Ghahreman, A., Ghahremani-nejad, F. and Matin, F. 1999.*Astragalus gigantirostratus* (Fabaceae), a remarkable new species from N Iran and supplementary notes on *A.* sect. *Cytisodes* Bunge. Willdenowia **29**: 221-225.

Maassoumi, A.A. 2000. The genus *Astragalus* in Iran, Perennials. Research Institute of Forests and Rangeland Publications, Tehran, Iran, pp. 441.

Maassoumi, A.A. 2005. The Genus *Astragalus* in Iran. Research Institute of Forests and Rangeland Publications, Tehran, Iran, pp. 786.

Maassoumi, A.A.2007. Two new species of the genus *Astragalus* L. (Fabaceae) from Iran. Iranian Journal of Botany **13**(2): 78-81.

Nylander, J.A.A. 2004.MrModeltest, ver. 2. Program distributed by the author. Uppsala University Press, Uppsala, Sweden.

Podlech, D. 1986. Taxonomic and phytogeographical problems in *Astragalus* of the Old World and south-west Asia. Proceedings of the Royal Society of Edinburgh **89**: 37-43.

Podlech, D. 1999. New *Astragali* and *Oxytropis* from North Africa and Asia, including some new combinations and remarks on some species. Sendtnera **6**: 135-174.

Podlech, D. 2001. Contributions to the knowledge of the genus *Astragalus* L. (Leguminosae). Sendtnera **7**: 163-201.

Podlech, D. 2004. New species of *Astragalus* L. (Leguminosae), mainly from Iran. Annalen des Naturhistorischen Museums in Wien **105**: 565-596.

Podlech, D. and Sytin, A. 2010. Papilionaceae VI: *Astragalus* section *Ornithopodium*. *In*: Rechinger, K.H. (Ed.), Flora Iranica. Vol. **178**. Akademische Druck-u Verlagsanstalt, Wien, Austria, pp. 173-184.

Podlech, D., Zarre, SH., Maassoumi, A.A., Ekici, M. and Sytin, A. 2010. Papilionaceae VI: Astragaleae, *Astragalus* L. IV. *In*: Rechinger, K.H. (Ed.). Flora Iranica. Vol. **178**. Akademische druck-u verlagsanstalt, Wien, Austria, pp. 1-430.

Podlech, D. and Zarre, SH. 2013. A taxonomic revision of the genus *Astragalus* L. (Leguminosae) in the Old World. Naturhistorisches Museum, Wien, Austria, pp. 2439.

Posada, D. and Buckley, T. 2004. Model selection and model averaging in phylogenetics: advantages of Akaike information criterion and Bayesian approaches over likelihood ratio rates. Systematic Biology **53**: 793-808.

Ranjbar, M. 2004. *Astragalus* sect. *Dissitiflori* (Fabaceae) in Iran. Nordic Journal of Botany **24** (5): 523-531.

Ranjbar, M. and Karamian, R.2002. Taxonomic study of *Astragalus* sect. *Erioceras* (Fabaceae) in Iran, additional notes and key to the species. Nordic Journal of Botany **22**(6): 713-717.

Ronquist, F. and Huelsenbeck, J.P. 2003. MrBayes 3, Bayesian phylogenetic inference under mixed models. Bioinformatics **19**: 1572-1574.

Sanderson, M.J. and Liston, A. 1995. Molecularphylogenetic systematics of Galegeae, with special reference to *Astragalus. In*: Crisp, M. and Doyle, J.J.(Eds.). Advances in legume systematics. Vol. 7. Kew Publishing, London, UK, pp. 331-350.

Sanderson, M.J. and Wojciechowski, M.F. 1996. Diversification rates in a temperate legume clade: are there ''so many species'' of *Astragalus* (Fabaceae)? American Journal of Botany **83**: 1488-1502.

Sang, T., Crawford, D.J. and Stuessy, T.F. 1997. Chloroplast DNA phylogeny, reticulate evolution, and biogeography of *Paeonia* (Paeoniaceae). American Journal of Botany **84**: 1120-1136.

Seelanan, T., Schnabel, A. and Wendel, J.F. 1997. Congruence and consensus in the cotton tribe (Malvaceae). Systematic Botany **22**: 259-290.

Sheikh Akbari Mehr, R., Maassoumi, A. A., Saidi, A., Kazempour Osaloo, S. and Ghorbani Nohooji, M. 2012a.Morphological cladistic analysis of some bifurcate hairy sections of *Astragalus* (Fabaceae) in Iran. Turkish Journal of Botany **36**: 434-442.

Sheikh Akbari Mehr, R., Saidi, A., Kazempour Osaloo, S. and Maassoumi, A. A. 2012b. Phylogeny of *Astragalus* section *Dissitiflori* based on nrDNAITS and morphological data in Iran. Iranian Journal of Botany **18**: 1-9.

Stöver, B.C. and Müller, K.F. 2010. TreeGraph 2 Combining and visualizing evidence from different phylogenetic analyses. BMC Bioinformatics **11**: 7-15.

Swofford, D. L. 2002. PAUP*: Phylogenetic Analysis Using Parsimony (and other methods). Version. 4.0b10. Sinauer Associates Inc., Sunderland, Massachusetts, USA.

Tate, J.A. and Simpson, B.B. 2003. Paraphyly of *Tarasa* (Malvaceae) and diverse origins of the polyploid species.Systematic Botany **28**: 723-737.

Vural, C., Ekici, M., Akan, H. and Aytac, Z. 2008.Seed morphology and its systematic implications for genus *Astragalus* L. sections *Onobrychoidei* DC., *Uliginosi* Gray and *Ornithopodium* Bunge (Fabaceae). Plant Systematics and Evolution **274**: 255-263.

White, T.J., Bruns, T., Lee, S. and Taylor, J. 1990. Amplification and direct sequencing of fungal ribosomal RNA genes for phylogenetics. *In*: Innis, M., Gelfand, D., Sninsky, J. and White, T. (Eds). PCR protocols. A guide to methods and applications. Academic Press, San Diego, California, USA, pp. 315-322.

Wiens, J.J. 1998.Combining data sets with different phylogenetic histories. Systematic Biology **47**: 568-581.

Wojciechowski, M.F., Lavin, M. and Sanderson, M.J. 2004. A phylogeny of Legums (Leguminosae) based on analysis of the plastid *mat*K gene resolves many well-supported subclades within the family. American Journal of Botany **91**: 1846-1862.

Yoder, A.D., Irwin, J.A. and Payseur, B.A. 2001. Failure of the ILD to determine data combinability for slow loris phylogeny. Systematic Biology **50**(3): 408-424.

8

GENETIC VARIATION AND MOLECULAR RELATIONSHIPS AMONG EIGHT TAXA OF *DESMODIUM* DESV. BASED ON RAPD MARKERS

M. Oliur Rahman[1], Md. Zahidur Rahman, Sonia Khan Sony[2]
AND Mohammad Nurul Islam

Department of Botany, University of Dhaka, Dhaka-1000, Bangladesh

Keywords: Desmodium Desv.; RAPD; Genetic diversity; UPGMA; Bangladesh.

Abstract

Genetic variation and molecular relationships among eight taxa of *Desmodium* Desv. were assessed on the basis of random amplified polymorphic DNA (RAPD) markers. The banding patterns of eight taxa namely, *Desmodium gangeticum* (L.) DC., *D. heterocarpon* (L.) DC., *D. heterophyllum* (Willd.) DC., *D. motorium* (Houtt.) Merr., *D. pulchellum* (L.) Benth., *D. triflorum* (L.) DC., *D. triquetrum* (L.) DC. and *D. triquetrum* subsp. *alatum* (DC.) Prain were compared. A total of 81 DNA fragments were detected by 11 primers. Among the taxa studied *D. triquetrum* and *D. triquetrum* subsp. *alatum* were found to be most closely related followed by close proximity between *D. gangeticum* and *D. motorium*. The highest genetic distance was observed between *D. triflorum* and *D. heterophyllum* followed by *D. heterocarpon* and *D. heterophyllum*. UPGMA dendrogram was constructed to show the genetic relatedness among the taxa employed and the tree revealed a close proximity among *D. pulchellum*, *D. gangeticum* and *D. motorium*. In contrast, *D. heterophyllum* was found distantly related with rest of the taxa.

Introduction

Desmodium Desv. belongs to the family Fabaceae comprises about 280 species widespread in the tropical and subtropical regions (Puhua and Ohashi, 2010). In Bangladesh *Desmodium* is represented by 19 taxa (Ahmed *et al.*, 2009). They are annual to perennial herbs, undershrubs or shrubs, and characterized by possessing uni- or tri-foliolate leaves, simple raceme or panicle inflorescence and distinctly jointed pods. The systematics of the genus *Desmodium* is confusing and not yet resolved completely (Ohashi and Mill, 2000). Several taxonomic studies on *Desmodium* were carried out based on morphology and anatomy (Pedley and Rudd, 1996; Shaheeen, 2008; Puhua and Ohashi, 2010). Recently, Rahman and Rahman (2012) conducted a morphometric study of *Desmodium* and showed interspecific relationships among 14 species of the genus. However, molecular studies employing different DNA markers on this genus are very scanty (Yue *et al.*, 2010; Ahmad Haji *et al.*, 2016).

Recent progress in DNA marker technology have augmented the marker resources for genetic analyses of a wide variety of genomes. The development of random amplified polymorphic DNA (RAPD) markers generated by polymerase chain reaction (PCR) using arbitrary primers has resulted in alternative molecular markers for the detection of nuclear DNA polymorphism (Williams *et al.*, 1990). RAPD markers have application in many fields including DNA fingerprinting (Elavazhagan *et al.*, 2009), assessment of genetic diversity (Bodo Slotta and Porter, 2006), cultivar identification (Sipahi *et al.*, 2010), estimation of population genetics (Sales *et al.*,

---

[1]Corresponding author. Email: prof.oliurrahman@gmail.com
[2]Department of Botany, University of Barisal, Barisal 8200, Bangladesh

2001), hybridization (Caraway *et al.*, 2001), systematics (Vilatersana *et al.*, 2005), phylogeny reconstruction (Ahmed *et al.,* 2005), and genome mapping (Krutovaskii *et al.*, 1998).

In legume species, RAPD markers have proven to be a useful tool in studies analyzing genetic variation (Yamaguchi and Jabadi, 2004; Bisoyi *et al.*, 2010). Previous studies on *Desmodium* using isozymes were conducted mainly with species which are important as forage (Smith and Schaal, 1979; Imrie and Blogg, 1983). Application of RAPD markers for detecting genetic variation and interspecific relationships of *Desmodium* is very limited. Bedolla-Garcia and Lara-Cabrera (2006) applied RAPD markers to detect genetic variation within and among five population of *Desmodium sumichrastii* from Mexico. Very recently, Singh *et al.* (2016) employed RAPD analysis for DNA fingerprinting of only two species of *Desmodium*, *viz.*, *D. gangeticum* and *D. laxiflorum*. However, no detailed study based on RAPD markers for detecting genetic diversity and interspecific relationships in *Desmodium* was carried out so far. Therefore, the aims of the present study are two-fold: i) to detect the genetic diversity among eight *Desmodium* taxa, and ii) to infer the relationship among these taxa of *Desmodium* based on RAPD analysis.

Materials and Methods

Plant materials

Eight taxa of *Desmodium* were collected from different places of Bangladesh, *viz.*, *Desmodium gangeticum* (L.) DC., *D. heterocarpon* (L.) DC., *D. heterophyllum* (Willd.) DC., *D. motorium* (Houtt.) Merr., *D. pulchellum* (L.) Benth., *D. triflorum* (L.) DC., *D. triquetrum* (L.) DC. and *D. triquetrum* subsp. *alatum* (DC.) Prain (Table 1). Leaf samples were used for DNA isolation and were preserved at -80°C until further use. The voucher specimens are deposited at Dhaka University Salar Khan Herbarium (DUSH).

Table 1. List of *Desmodium* Desv. taxa used for RAPD analysis.

No.	Taxa	Habit	Voucher specimens
1.	*Desmodium gangeticum* (L.) DC.	Undershrub	Dhaka: 27.9.2011, Zahid 85 (DUSH); Munshigonj: 1.1.2011, Zahid 7 (DUSH).
2.	*D. heterocarpon* (L.) DC.	Undershrub	Cox's Bazar: Teknaf, 24.4.2011, Zahid 28 (DUSH); Gazipur: Rajendrapur, 5.11.2011, Zahid 91(DUSH).
3.	*D. heterophyllum* (Willd.) DC.	Herb	Gazipur: Rajendrapur, 15.7.2011, Zahid 50 (DUSH); Cox's Bazar: Pekua, 21.8.2011, Zahid 74 (DUSH).
4.	*D. motorium* (Houtt.) Merr.	Undershrub	Dhaka: 23.12.2010, Zahid 02 (DUSH).
5.	*D. pulchellum* (L.) Benth.	Shrub	Gazipur: Rajendrapur, 15.7.2011, Zahid 48 (DUSH).
6.	*D. triflorum* (L.) DC.	Herb	Cox's Bazar: Kutubdia island, 17.7.2011, Zahid 63 (DUSH); Narsingdi: Wari Boteshwar, 10.11.2011, Zahid 96 (DUSH).
7.	*D. triquetrum* (L.) DC.	Shrub	Cox's Bazar: Teknaf, 24.4.2011, Zahid 31(DUSH).
8.	*D. triquetrum* subsp. *alatum* (DC.) Prain	Shrub	Cox's Bazar: Teknaf, 19.7.2011, Zahid 73 (DUSH).

Genomic DNA isolation

DNA was isolated from leaves using the CTAB (Cetyl trimethyl ammonium bromide) method following Doyle and Doyle (1987). The isolated DNA was preserved in TE buffer and stored at −20°C.

RAPD amplification

The oligonucleotide primers tested for RAPD analysis were presented in Table 2. These primers were chosen by their number and consistency of amplified fragments for analyzing *Desmodium* taxa. The amplification reaction contained 50 ng of genomic DNA, 0.5 unit of Taq DNA polymerase, 0.5 µl of each dNTPs, 10 mM $MgCl_2$, 1µl decamer random primers (Operon Biotechnology, Germany) and 2.5 µl 10X amplification buffer in a total volume of 25 µl. The amplifications were performed in triplicate using PCR thermal cycler (Biometra UNOII, Germany) with initial denaturation of 5 min at 94°C, followed by 42 cycles at 94°C for 5 sec, 33°C for 1 min and 72°C for 2 min with final extension of 5 min at 72°C. The amplified products were separated on 1% agarose gel containing ethidium bromide, and photographed under UV light.

Table 2. List of primers used in RAPD analysis.

Primer Code	Sequence (5′-3′)	G + C content (%)
OPA-1	TGCCGAGCTC	70
OPA-2	TGCCGAGCTG	70
OPA-3	AGTCAGCCAC	60
OPA-6	GGTCCCTGAC	70
OPA-7	GAAACGGGTG	60
OPA-8	GTGACGTAGG	60
OPA-9	GTGATCGCAG	60
OPA-10	GTGATCGCAG	60
A15	TTCCGAACCC	60
B14	TCCGCTCTGG	70
BO6	TGCTCTGCCC	70

Data analysis

RAPD bands were recorded in a binary data matrix scored as presence (1) or absence (0). The score obtained using all primers in the RAPD analysis were then combined to create a single data matrix. The size of amplification products were estimated by comparing the migration of each amplified fragments with that of a known size fragments of 1 kb molecular weight marker. Genetic linkage distance was determined using the data matrix. UPGMA (Unweighted pair group method with arithmetic means) dendrogram was constructed to show the genetic relationships among the species (Sneath and Sokal, 1973). All analyses were performed using the Statistica program.

Results and Discussion

A total of 81 RAPD bands were scored with eleven RAPD primers in eight *Desmodium* taxa. The highest number of fragments was detected in *Desmodium heterophyllum* (31) followed by *D. triflorum* (20) and *D. heterocarpon* (7), while the lowest band observed in *D. motorium* (1). The highest number of fingerprints were generated by the primer OPA-8 and least number in OPA-1.

The RAPD markers have been found efficient to detect genetic variation in *Desmodium*. The highest dissimilarity (41.0) was observed between *D. triflorum* and *D. heterophyllum* followed by *D. pulchellum* and *D. heterophyllum* (35.5) and *D. heterophyllum* and *D. heterocarpon* (35.0) (Table 3). The lowest genetic distance (1.0) was found between *D. gangeticum* and *D. motorium* indicating that these species are very closely related (Table 3).

Cluster analysis of the genetic similarity estimates from RAPD markers was performed to generate the UPGMA dendrogram for showing genetic relationship among the taxa of *Desmodium* (Fig. 1). The dendrogram revealed that *D. pulchellum, D. gangeticum, D. motorium, D. triquetrum, D. triquetrum* subsp. *alatum,* and *D. heterocarpon* grouped together and formed a cluster showing a close relationships among them. This cluster further consisted of two sub-clusters, the first one contained *D. pulchellum, D. gangeticum* and *D. motorium* showing a close affinity between these three species, while the second sub-cluster comprised *D. triquetrum* subsp. *alatum, D. triquetrum,* and *D. heterocarpon*. The highest relatedness was observed between *D. gangeticum* and *D. motorium* among all the taxa employed in this study. The RAPD analysisalso shown that *D. heterophyllum* and *D. triflorum* retained ungrouped and they are distantly related from other taxa of *Desmodium*.

Table 3. Genetic variation among studied taxa of *Desmodium*.

Taxa	*D. pulche-llum*	*D. triflo-rum*	*D. hetero-carpon*	*D. hetero-phyllum*	*D. triquetrum* subsp. *alatum*	*D. trique-trum*	*D. gange-ticum*	*D. moto-rium*
D. pulchellum	0							
D. triflorum	28.3	0						
D. heterocarpon	14.1	32.2	0					
D. heterophyllum	35.5	41.0	35.0	0				
D. triquetrum subsp. *alatum*	8.0	26.3	14.1	33.5	0			
D. triquetrum	8.0	26.3	12.1	33.5	8.0	0		
D. gangeticum	6.0	24.3	12.1	33.6	6.0	6.0	0	
D. motorium	5.0	25.4	11.1	32.6	5.0	5.0	1.0	0

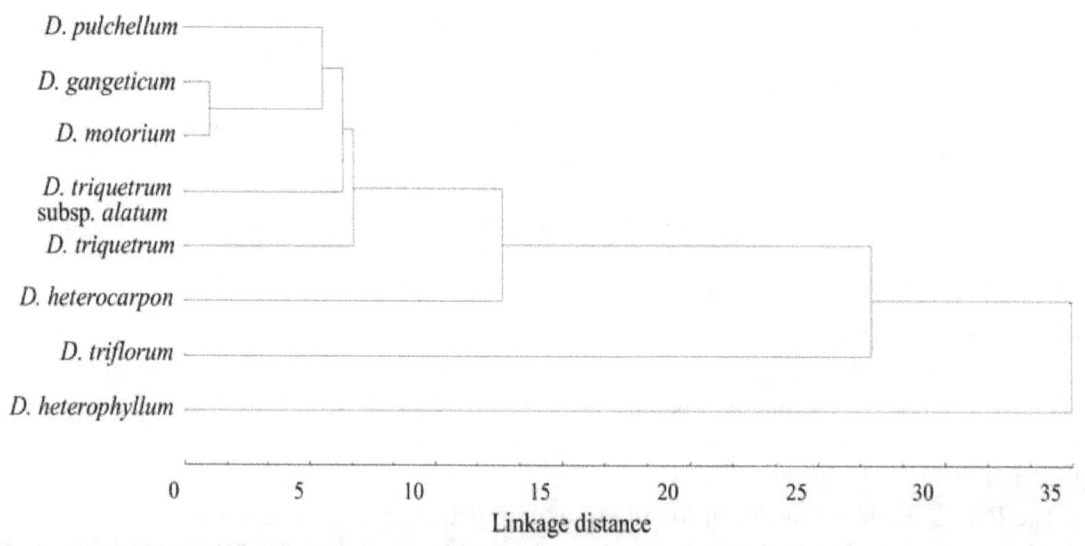

Fig. 1. UPGMA dendrogram showing the genetic relationship among studied Desmodium taxa based on RAPD markers.

The RAPD method is popular because of its technical simplicity and speed. The present study reveals that RAPDs are useful markers in identifying *Desmodium* species. The RAPD data shows that *D. gangeticum* and *D. motorium* are genetically closely related. This result is congruent with the previous study based on morphological characters such as unifoliolate, lanceolate leaves and triangular stipules (Ahmed *et al.,* 2009); and also supported by foliar anatomical investigation (data not shown). Bedolla-Garcia and Lara-Cabrera (2006) studied genetic variation within and among five populations of *Desmodium sumichrastii* from Mexico based on RAPD analysis. Singh *et al.* (2016) employed RAPD approach for the genetic fingerprinting of *Desmodium gangeticum* and *D. laxiflorum* and found 60-65% similarity between these two species. Irshad *et al.* (2009) studied three species of *Desmodium,* viz. *D. gangeticum, D. triflorum* and *D. velutinum* (Willd.) DC. and compared with commercial samples of various origin. Among these *D. triflorum* appears closer to *D. gangeticum* reflecting narrow genetic diversity. However, the present study shows that *D. triflorum* and *D. gangeticum* are distantly related. Very recently, Malgaonkar *et al.* (2016) determined the genetic relatedness and diversity among accessions of four *Desmodium* species using RAPD markers, namely *D. dichotomum* (Willd.) DC., *D. laxiflorum* DC., *D. scorpiurus* (SW.) Poir. and *D. triflorum.* A close affinity has been observed between *D. laxiflorum* and *D. scorpiurus.* In the present study *D. pulchellum* has been found close to *D. gangeticum* and *D. motorium* indicating that these three species are closely related. *D. triflorum* and *D. heterophyllum* are closely allied as evidenced by anatomical study (Data not shown), however, this affinity is not supported by RAPD analysis. In order to have better understanding about genetic relatedness and interspecific relationships inclusion of more taxa with additional markers is necessary.

Acknowledgements

The first author gratefully acknowledges the financial assistance provided by the Biotechnology Research Centre of the University of Dhaka for carrying out the research. The authors are grateful to Prof. Rakha Hari Sarkar and Prof. Md. Imdadul Hoque, Department of Botany, University of Dhaka for their cooperation during the course of the study. Thanks are also due to Dr. Sujay Kumar Bhajan of the Department of Botany, University of Dhaka for his help and cooperation.

References

Ahmad Haji, R.F., Tiwari, S., Gandhi, S.G., Kumar, A., Brindavanam, N.B. and Verma, V. 2016. Genetic diversity analysis among accessions of *Desmodium gangeticum* (L.) DC. with simple sequence repeat (SSR) and internal transcribed spacer (ITS) regions for species conservation. J. Biodivers. Biopros. Dev. **3**: 159.

Ahmed, S.M., Verma, V., Qazi, P.H., Ganaie, M.M., Bakshi, S.K. and Qazi, G.N. 2005. Molecular phylogeny in Indian *Tinospora* species by DNA based molecular markers. Plant Syst. Evol. **256**(1-4): 75–87.

Ahmed, Z.U., Hassan, M.A., Begum, Z.N.T. Khondoker, M., Kabir, S.M.H., Ahmed, M., Ahmed, A.T.A., Rahman, A.K.A. and Haque, E.U. (Eds) 2009. Encyclopedia of Flora and Fauna of Bangladesh, Vol. **8**. Angiosperms: Dicotyledons (Fabaceae - Lythraceae). Asiatic Society of Bangladesh, Dhaka. pp. 1–478.

Bedolla-Garcia, B.Y. and Lara-Cabrera, S.I. 2006. An assessment of genetic diversity in *Desmodium sumichrastii* (Fabaceae) in Mexico. Can. J. Bot. **84**: 876–882.

Bisoyi, M.K., Acharya, L., Mukherjee, A.K. and Panda, P.C. 2010. Study of inter-specific relationship in six species of *Sesbania* Scop. (Leguminosae) through RAPD and ISSR markers. International J. Plant Physiol. & Biochem. **2**(2): 11–17.

Bodo Slotta, T.A. and Porter, D.M. 2006. Genetic variation within and between *Iliamna corei* and *I. remota* (Malvaceae): implication for species delimitation. Bot. J. Linn. Soc. **151**: 345–354.

Caraway, V., Carr, G.D. and Morden, C.W. 2001. Assessment of hybridization and introgession in lava colonizing Hawaiian *Dubautia* (Asteraceae: Madiinae) using RAPD markers. Am. J. Bot. **88**: 1688–1694.

Doyle, J.J. and Doyle, J.L. 1987. A rapid DNA isolation procedure for small quantities of fresh leaf tissue. Phytochem. Bull. **19**: 11–15.

Elavazhagan, T., Ramakrishnan, M., Jayakumar, S., Chitravadivu, C. and Balakrishnan, V. 2009. Fingerprinting analysis in *Casurina equisetifolia* by using RAPD markers. Botany Res. Int. **2**(4): 244–247.

Imrie, B.C. and Blogg, D. 1983. Variability in isozyme gene frequency in the tropical pasture legumes, 'Greenleaf' *Desmodium.* Trop. Agric. **60**: 193–196.

Irshad, S., Singh, J., Kakkar, P. and Mehrotra, S. 2009. Molecular characterization of *Desmodium* species – An important ingredient of 'Dashmoola' by RAPD analysis. Fitoterapia **80**: 115–118.

Krutovskii, K.V., Vollmer, S.S., Sorensen, F.C., Adams, W.T., Knapp, S.J. and Strauss, S.H. 1998. RAPD genome map of Douglas-fir. J. Heredity **89**: 197–207.

Malgaonkar, M., Murthy, S.N. and. Pawar, S.D. 2016. Molecular analysis and study of genetic relationships among species of *Desmodium* Desv. using RAPD markers. J. Adv. Biol. & Biotech. **6**(4): 1–7.

Ohashi, H. and Mill, R.R. 2000. *Hylodesmum*, a new name for *Podocarpium* (Leguminosae). Edinburgh J. Bot. **57**: 171–188.

Pedley, L. and Rudd, V.E. 1996. Fabaceae, Tribe Desmodieae. *In:* Dassanayake, M.D. and Clayton, W.D. (Eds), Flora of Ceylon. Vol. **10**. Oxford & IBH publishing Co. Pvt. Ltd., New Delhi. pp. 149–198.

Puhua, H. and Ohashi, H. 2010. *Desmodium* Desv. *In*: Wu, Z.Y., Raven, P.H. and Hong, D.Y. (Eds), Flora of China. Vol. **10** (Fabaceae). Science Press, Beijing, and Missouri Botanical Garden Press, St. Louis. pp. 268–278.

Rahman, M.Z. and Rahman, M.O. 2012. Morphometric analysis of *Desmodium* Desv. (Fabaceae) in Bangladesh. Bangladesh J. Bot. **41**(2): 143–148.

Sales, E., Nebauer, S.G., Mus, M. and Segura, J. 2001. Population genetics study in the Balearic endemic plant species *Digitalis minor* (Scrophulariaceae) using RAPD markers. Am. J. Bot. **88**: 1750–1759.

Shaheen, A.S.M. 2008. Morphological and anatomical investigation in *Desmodium tortuosum* (SW.) DC. (Fabaceae): A new addition to the Egyptian flora. Bangladesh J. Plant Taxon. **15**(1): 21–29.

Singh, S., Harisha C.R., Goyal, M. and Patel, B.R. 2016. Comparative molecular characterization of *Desmodium gangeticum* DC. and *Desmodium laxiflorum* DC. through random amplified polymorphic DNA (RAPD) analysis. Asian J. Plant Sci. & Res. **6**(1): 22–26.

Sipahi, H., Akar, T., Yildiz, M.A. and Sayim, I. 2010. Determination of genetic variation and relationship in Turkish Barley cultivars by hordein and RAPD markers. Turkish J. Field Crops **15**(2): 108–113.

Smith, W.G. and Schaal, B.A. 1979. Isozme variation in *Desmodium nudiflorum*. Biochem. Syst. Ecol. **7**: 121–123.

Sneath, P.H.A. and. Sokal, R.R. 1973. Numerical Taxonomy. Freeman and Company, San Francisco. 573 pp.

Vilatersana, R., Garnatje, T., Susanna, A. and Garcia-Jacas, N. 2005. Taxonomic problems in *Carthamus* (Asteraceae): RAPD markers and sectional classification. Bot. J. Linn. Soc. **147**: 375–383.

Williams, J.G.K., Kubelik, A.R., Livak, K.J., Rafalski, J.A. and Tingey, S.V. 1990. DNA polymorphisms amplified by arbitrary primers are useful as genetic markers. Nucleic Acids. Res. **18**: 176–183.

Yamaguchi, H. and Jabadi, F. 2004. RAPD and seed coat morphology variation in annual and perennial species of the genus *Cicer* L. Plant Genet. Res. Crop Evol. **51**: 783–794.

Yue, M.-F., Zhou, R.-C., Huang, Y.-L., Xin, G.-R., Shi, S.-H. and Feng, L. 2010. Genetic diversity and geographical differentiation of *Desmodium triflorum* (L.) DC. in south China revealed by AFLP markers. J. Plant Biol. **53**: 165–171.

9

MOLECULAR PHYLOGENETIC ANALYSES OF INTERNAL TRANSCRIBED SPACER (ITS) SEQUENCES OF NUCLEAR RIBOSOMAL DNA INDICATE MONOPHYLY OF THE GENUS *PHYTOLACCA* L. (PHYTOLACCACEAE)

M. Ajmal Ali[1], Joongku Lee[2], Soo-Yong Kim[2], Sang-Hong Park[2,3]
AND Fahad M.A. Al-Hemaid

Department of Botany and Microbiology, College of Science, King Saud University, Riyadh 11451, Kingdom of Saudi Arabia

Keywords: ITS; nrDNA; Phytolaccaceae; Phylogeny.

Abstract

Relationships within the family Phytolaccaceae *sensu lato* were examined based on internal transcribed spacer (ITS) sequences of nuclear ribosomal DNA (nrDNA). The study revealed *Phytolacca* L. as taxonomically the most difficult genus in the family with completely unknown phylogeny. Molecular evidence was used from nrDNA ITS sequences of about 90% of the species for maximum parsimony analyses, and the molecular phylogenetic analyses defined a monophyletic *Phytolacca*. This first molecular phylogenetic study of *Phytolacca* concludes that the relationships among the species within the genus do not show harmony with the generic classification based on morphology. These results set the stage for a more detailed phylogenetic analysis of *Phytolacca*.

Introduction

The angiosperm family Phytolaccaceae *sensu lato* comprises a weedy, and polyphyletic genera (APGIII, 2009) of largely tropical and subtropical plants that have been placed, almost without exception, in Centrospermae under either the order Chenopodiales or Caryophyllales (Nowicke, 1969). The genus *Phytolacca* L. (family Phytolaccaceae) is commonly known as '*pokeweeds*' comprises about 20 species (Nowicke, 1969) of perennial herbs, shrubs and trees, nearly cosmopolitan, mostly native to South America, with a few species in Africa and Asia (Shu, 2003). The genus *Phytolacca* possess alternate, simple leaves, pointed at the end, with entire or crinkled margins; the leaves can be either deciduous or evergreen; the stems are green, pink or red; the flowers are greenish-white to pink, produced in long racemes at the ends of the stems; they develop into globose berries 4–12 mm in diameter, green at first but dark purple to black after ripening (Nowicke, 1969).

The generic name is derived from the Greek word *phyton*, meaning plant, and the Latin word *lacca*, meaning a red dye (Umberto, 2000). Phytolaccatoxin and phytolaccigenin, which are poisonous, are present in many species of the genus *Phytolacca*. The active principles for analgesic, anti-inflammatory, bactericidal, fungicidal, mitogenic and molluscicide action have been reported from several species of *Phytolacca* (Hernández *et al.*, 2013). The active principles have also been found in methanolic extracts of fruit of *P. tetramera* Hauman, which is a source of saponins with fungicidal action (Escalante *et al.*, 2002; Santecchia *et al.*, 2002). The African soapberry plant, *P. dodecandra* L'Her., locally called *endod*, produces a range of triterpenoid

---
[1]Corresponding author. Email: alimohammad@ksu.edu.sa
[2]International Biological Material Research Center, Korea Research Institute of Bioscience and Biotechnology, 125 Gwahak-ro, Yuseong-gu, Daejeon 305-806, South Korea. Email: joongku@kribb.re.kr
[3]Present address: Division of Plant Management, National Institute of Ecology, Choongnam, Secheon-gun, Maseo-myeon, Geumgang-ro, 1210, 325-813, South Korea

saponins possessing very potent and useful biological properties, including antifungal, anti-protozoan, spermicidal and insecticidal activities (Lemma *et al.*, 1979). Because of its fast-growing nature, *P. dioica* L. is frequently planted as a shade tree in the tropics. Nowicke (1969) reported the use of berries and the young sprouts, and leaves of some species of *Phytolacca* as an adulterant of red wine and poke salad, respectively.

The generic composition and phylogeny of Phytolaccaceae have long been controversial. The phylogenetic studies have substantially added new results to our knowledge of phylogeny of the family Phytolaccaceae (Brown and Varadarajan, 1985; Downie *et al.*, 1997; Cuenoud *et al.*, 2002; Lee *et al.*, 2013). Nowicke (1969) referred *Phytolacca* as the most difficult genus in the family Phytolaccaceae *sensu lato*, and classified under three subgenera and six sections (Table 1). However, comprehensive information on phylogeny of the genus *Phytolacca* is lacking.

Table 1. Infrageneric classification of the genus *Phytolacca* L. by Nowicke (1969). Taxa included in the present study are marked with asterisk.

Subgenus	Section	Species
Pircunia	*Pircunia*	**Phytolacca acinosa* Roxb.
		*P. heptandra Retz.
	Pircunioides	**P. dodecandra* L'Her.
Pircuniopsis	*Pircuniophorum*	**P. sanguinea* H. Walter
		*P. rugosa Br. & Bouche
		P. chilensis (Miers *ex* Moq.) H. Walter
	Pircuniopsis	**P. tetramera* Hauman
		*P. dioica L.
		*P. weberbaueri H. Walter
Phytolacca	*Phytolacca*	**P. icosandra* L.
		*P. octandra L.
		*P. thyrsiflora Fenzl *ex* J.A. Schmidt
		*P. heterotepala H. Walter
		*P. meziana H. Walter
		*P. rivinoides Kunth & Bouchk
		*P. purpurascens A. Br. & Bouche
		*P. brachystachys Moq.
		*P. bogotensis H.B.K.
		*P. americana L.
	Phytolaccoides	*P. pruinosa* Fenzl.

During the last two decades, the internal transcribed spacers (ITS) sequences of nuclear ribosomal DNA (nrDNA) have gained wide attention, not only because of its efficacy in understanding phylogeny of the plants at lower taxonomic level, but also to be considered as the most conserved markers, because, even after facing criticism of its utility, this marker stands parallel to the smartest genes available for the molecular phylogeny and plant DNA barcoding (Ali *et al.*, 2013, 2014). The nrDNA ITS sequences have, therefore, provided a useful source of phylogenetic information in many genera and families of flowering as well as non-flowering plants (Ali *et al.*, 2015), including Phytolaccaceae (Lee *et al.*, 2013). Hence, as such the nrDNA ITS are appropriate to analyze for the genus *Phytolacca* too.

Materials and Methods

Taxa examined

Twenty taxa representing five sections (i.e. *Phytolacca, Pircunia, Pircunioides, Pircuniophorum* and *Pircuniopsis*) under three subgenera (i.e. *Phytolacca, Pircunia* and *Pircuniopsis*) of *Phytolacca* and two outgroup taxa (namely *Petiveria alliacea* F. Muell. and *Monococcus echinophorus* L.) were sampled from specimens deposited in the Herbarium of University of California (UC), Berkeley, USA (Table 2). *Petiveria alliacea* and *M. echinophorus* were chosen as outgroup taxa because of their close affinity to *Phytolacca* (Lee *et al.*, 2013).

Table 2. Accessions of the genus *Phytolacca* L. examined in this study.

Taxon	Voucher	Locality	GenBank Acc. No.
Ingroup			
Phytolacca acinosa	*M.T. Yu et al. s.n.*	Tibet	EU239681
P. americana	*D.W. Taylor 7922* (UC/JEPS)	California, USA	JX232573
P. bogotensis	*H.L. Mason 23712* (UC)	Colombia, South America	KM491868
P. brachystachys	*F.R. Fosberg 9004* (UC)	Hawaiian Island, USA	KM491869
P. dioica	*Marquez et al. 38645* (UC)	Mexico, North America	JX232571
P. dodecandra	*R.E.S. Tanner 572* (UC)	Tanganyika, Africa	KM491870
P. heptandra	*L.C.C. Libeoberg 5830* (UC)	South Africa	KM491871
P. heterotepala	*Sally Pugh s.n.* (UC)	California, USA	KM491872
P. icosandra	*J.H. Beaman 2749* (UC)	Mexico, North America	JX232570
P. meziana	*Edward 89055* (UC)	Mexico, North America	KM491873
P. octandra	*G.J. Martin 468* (UC)	Oaxaca, Mexico, North America	KM491874
P. purpurascens	*W.H. Wagher 5027* (UC)	Hawaiian Island, USA	KM491875
P. rivinoides	*J. Nowicke 874* (UC)	Panama, Central America	KM491876
P. rugosa	*A. Weston 5981* (UC)	Costa Rica, Central America	KM491877
P. sanguinea	*J. H. Langenneim 3576* (UC)	Colombia, South America	KM491878
P. tetramera	*N. Tur 1329* (UC)	Argentina, South America	KM491879
P. thyrsiflora	*C. Chung 4248* (UC)	California, USA	KM491880
P. weberbaueri	*C.H. Dodson 6481* (UC)	Ecuador, South America	KM491881
Outgroup			
Monococcus echinophorus	*Franch 1130* (UC)	New Caledonia	JX232579
Petiveria alliacea	*C.A. Purpus 2272* (UC)	Mexico, North America	JX232580

Molecular methods

Total genomic DNA was extracted by use of the DNeasy Plant Mini Kit from Qiagen (Valencia, CA, USA). The nrDNA ITS regions were amplified using the primers ITS1 and ITS4 (White *et al.*, 1990). The DNA amplification for 35 cycles was carried out through PCR. Initial denaturation was carried out at 94°C for 5 min, followed by denaturation at 94°C for 1 min, annealing at 48°C for 1 min, extension at 72°C for 1 min, and the final extension at 72°C for 5 min. The PCR products were purified using SolGent PCR Purification kit-Ultra (SolGent, Daejeon, South Korea). For sequencing, the Big Dye Terminator chemistry (ABI) and an ABI 3100 Avant capillary sequencer were used. All sequences were BLAST-searched in GenBank.

Sequence alignments and phylogenetic analyses

Sequences were edited using the ABI Sequence Navigator (Perkin-Elmer/Applied Biosystems, USA). Sequence alignment was performed using Clustal X version 1.81 (Thompson *et al.*, 1997), and subsequently adjusted manually using BioEdit (Hall, 1999). Information on sequence alignment can be made available from the corresponding author. Data were exported as a nexus file and subsequently analyzed using Maximum Parsimony (MP) in PAUP* 4.0b10 (Swofford, 2002). The MP analysis was performed with the following settings: heuristic search algorithms with tree bisection reconnecting (TBR) branch swapping, MULPARS in effect, all characters equally weighted, gap treated as missing characters, zero-length branches collapsed, random addition sequence set to 1000 replicates, and branch swapping limited to 10,000,000 rearrangements per replicate. When maximum parsimony trees were saved, a strict consensus tree was constructed. Bootstrap analysis was performed using 1000 replicates, with the random addition sequence set to 10, and branch swapping limited to 10,000,000 rearrangements per replicate.

Results and Discussion

Sequence characteristics

The combined length of the entire ITS region (ITS1, 5.8S and ITS2) from taxa analyzed in the present study ranged from 609–631 nucleotides (nt). The length of the ITS1 region and GC contents ranged from 220–232 nt and 56%–63%, the 5.8S gene was 166 nt long, the length of the ITS2 region and the GC content ranged from 221–235 nt and 55%–63%, respectively. Data matrix has a total number of 654 nt characters of which 423 nt characters were constant, 88 nt characters were variable but parsimony-uninformative, and 143 nt characters were parsimony-informative.

Phylogenetic analyses

The parsimony analysis of the entire ITS region resulted a total number of four maximally parsimonious trees (MPTs) with a total length of 252 steps, a consistency index (CI) of 0.7110, a homoplasy index (HI) of 0.2890, rescaled consistency index (RC) of 0.5361 and a retention index (RI) of 0.7540 (Fig. 1).

The rooted bootstrap strict consensus parsimony tree (Fig. 1) revealed that the monophyly of *Phytolacca* species is supported with 100% parsimony bootstrap support (BS). All trees resulted from the analysis of ITS sequences resolve three major clades (Clades I–III, Fig. 1). The Clade I consists of *P. heptandra*, the Clade II (96% BS) consists of members of subgenus *Pircuniopsis* (i.e. *P. dioica*, *P. tetramera* and *P. weberbaueri*), and the Clade III (56% BS) consists of [*P. americana* + (*P. dodecandra* - *P. acinosa* - *P. purpurascens*) + *(P. rivinoides* - {*P. rugosa* - *P. thyrsiflora* + *P. icosandra* - *P. brachystachys* - *P. heterotepala* + *P. octandra* - *P. meziana* - *P. sanguinea* - *P. bogotensis*})].

The generic composition of *Phytolacca* has long been controversial principally due to common occurrence of intraspecific variability and hybridization (Fassett and Sauer, 1950; Sauer, 1951). Walter (1909) placed 26 species of *Phytolacca* into three subgenera based on the degree of connation of the carpels: free, connate at the base with the apices free, or completely united carpels. The subgenus *Pircunia* (Moq.) H. Walter contains *P. heptandra* Retz., *P. esculenta* van Houtte, *P. acinosa* Roxb., *P. latbenia* (Buch.-Ham.) H. Waiter and *P. cyclopetala* H. Walter under the Sect. *Pircuniastrum* Moq. characterized by hermaphroditic flowers, and *P. dodecandra*, *P. goudotii* Briq. and *P. nutans* H. Walter under the Sect. *Pircunioides* H. Walter characterized by dioecious plants. The subgenus *Pircuniopsis* H. Walter characterized by carpels connate at the base with the apices free, contains a hermaphroditic group, the Sect. *Pircuniophorum* H. Walter, with three species, *P. chilensis* (Miers *ex* Moq.) H. Walter, *P. rugosa* Br. & Bouche and *P. sanguinea* H. Walter, and the Sect. *Pseudolacca* Moq., with two dioecious species, *P. dioica* and *P. weberbaueri* H. Walter. The subgenus *Euphytolacca* Moq., the largest group characterized by carpels completely united contains a very large hermaphroditic flower, has the Sect. *Phytolaccastrum* H. Walter with *P. americana* L., *P. australis* Phil., *P. brachystachys* Moq., *P. heterotepala* H. Walter, *P. icosandra* L., *P. meziana* H. Walter, *P. micrantha* H. Walter, *P. octandra* L., *P. polyandra* Batalin, *P. purpurascens* A. Br. & Bouche, *P. rivinoides* Kunth & Bouchk and *P. thyrsiflora* Fenzl *ex* J.A. Schmidt, and a monotypic dioecious Sect. *Phytolaccoides* H. Walter containing *P. pruinosa* Fenzl. Later on Heimerl (1934) noted approximately 35 species of *Phytolacca*; however, Nowicke (1969) did not consider the names assigned to hybrid origin. Nowicke (1969) recognized a total of 20 species in the genus *Phytolacca* and classified them into

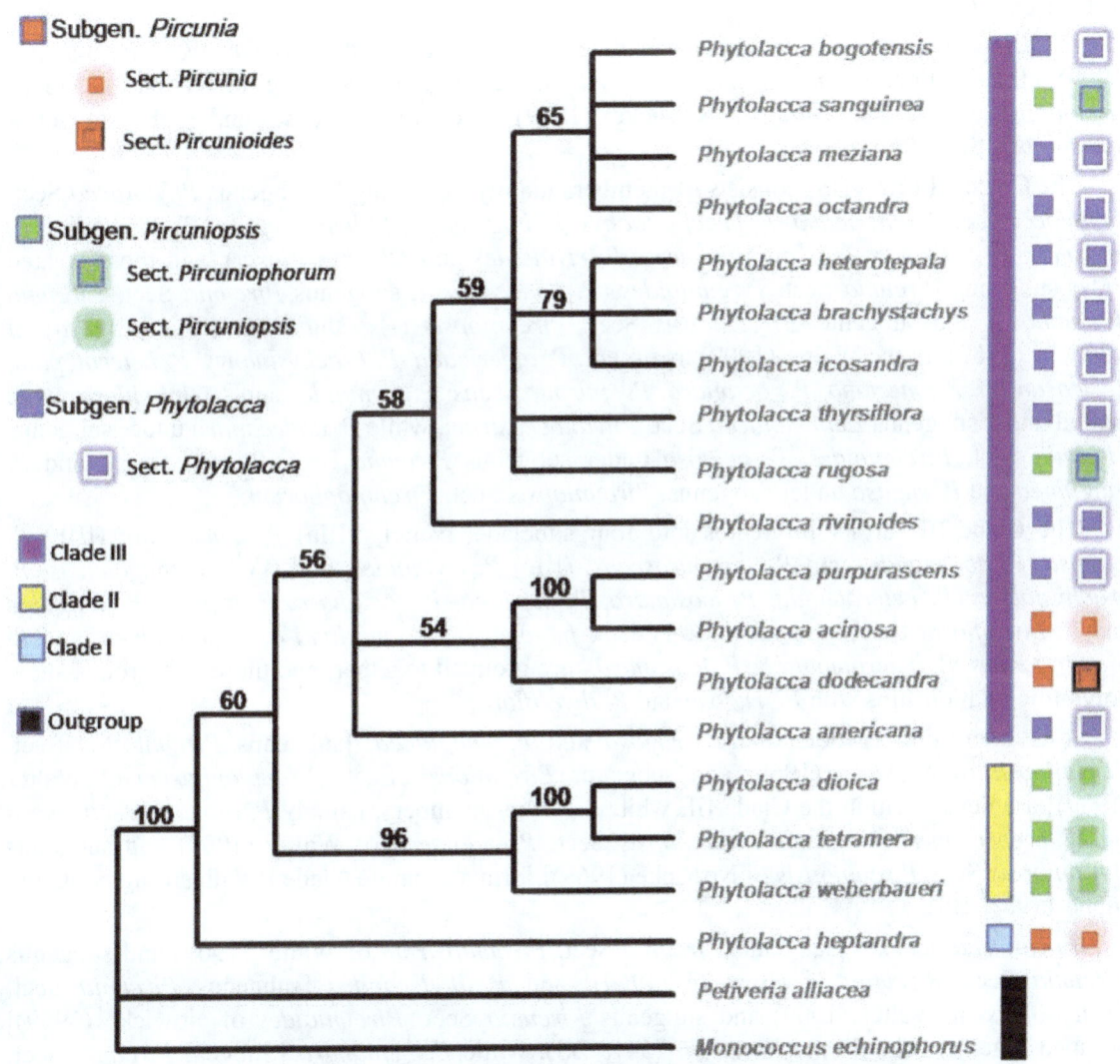

Fig. 1. The bootstrap strict consensus of four maximally parsimonious trees of *Phytolacca* L. species based on the ITS sequence with gaps being treated as missing data (252 steps, CI= 0.71, HI= 0.28, RC= 0.53 and RI= 0.75). Bootstrap values greater than 50% in 1000 replicates are shown above lines.

three subgenera, i.e. *Pircunia* (carpels completely free), *Pircuniopsis* (carpels more or less united) and *Phytolacca* (carpels completely united, the styles more or less connivent). Based on characteristic of flowers, Nowicke (1969) divided the subgenus *Pircunia* into two sections: *Pircunia* (*P. acinosa* and *P. heptandra*) and *Pircunioides* (*P. dodecandra*); *Pircuniopsis* into two sections: *Pircuniophorum* (*P. chilensis*, *P. rugosa* and *P. sanguinea*) and *Pircuniopsis* (*P. dioica*, *P. tetramera* and *P. weberbaueri*); and *Phytolacca* into two sections: *Phytolacca* (*P. americana*, *P. bogotensis* H.B.K., *P. brachystachys*, *P. heterotepala*, *P. icosandra*, *P. meziana*, *P. octandra*, *P. purpurascens*, *P. rivinoides* and *P. thyrsiflora*) and *Phytolaccoides* (*P. pruinosa*).

In our study, the Clade I, which occupies independently at the basal position in MPT, consists of only *P. heptandra*. *Phytolacca heptandra* was treated along with *P. esculenta*, *P. acinosa*, *P. latbenia* and *P. cyclopetala* under the subgenus *Pircunia*, Sect. *Pircuniastrum* (Walter, 1909). Nowicke (1969) also treated *P. heptandra* along with *P. acinosa* under the subgenus *Pircunia* Sect. *Pircunia*.

The Clade II (96% BS) consists of *P. dioica*, *P. tetramera* and *P. weberbaueri*. In Walter's (1909) classification *P. dioica*, *P. tetramera* and *P. weberbaueri* are under the subgenus *Pircuniopsis*, Sect. *Pseudolacca*. Nowicke (1969) also treated these under the subgenus *Pircuniopsis* Sect. *Pircuniopsis*.

The Clade III (56% BS) consists of members mainly belonging to subgenus *Phytolacca* Sect. *Phytolacca* (i.e. *P. americana*, *P. brachystachys*, *P. bogotensis*, *P. heterotepala*, *P. icosandra*, *P. meziana*, *P. octandra*, *P. purpurascens*, *P. rivinoides* and *P. thyrsiflora*,), and those treated under subgenus *Pircunia* Sect. *Pircunioides* (*P. dodecandra*), subgenus *Pircunia* Sect. *Pircunia* (*P. acinosa*) and subgenus *Pircuniopsis* Sect. *Pircuniopsis* (*P. sanguinea* and *P. rugosa*) of Nowicke (1969). In the Walter (1909) treatment, *P. americana*, *P. brachystachys*, *P. heterotepala*, *P. icosandra*, *P. meziana*, *P. octandra*, *P. purpurascens*, *P. rivinoides* and *P. thyrsiflora*, were treated under subgenus *Euphytolacca* Sect. *Phytolaccastrum*, while *P. dodecandra* under subgenus *Pircunia* Sect. *Pircunioides*, *P. acinosa* under subgenus *Pircunia* Sect. *Pircuniastrum*, and *P. sanguinea* and *P. rugosa* under subgenus *Pircuniopsis* Sect. *Pircuniophorum*.

The Clade III further bifurcates into four subclade, namely (IIIa) *P. americana*; (IIIb) *P. acinosa*, *P. dodecandra* and *P. purpurascens*; (IIIc) *P. rivinoides*; and (IVd) *P. bogotensis*, *P. brachystachys*, *P. heterotepala*, *P. icosandra*, *P. meziana*, *P. octandra*, *P. rugosa*, *P. sanguine* and *P. thyrsiflora*. Under the subclade IVd, *P. bogotensis*, *P. meziana* & *P. sanguinea*, and *P. brachystachys*, *P. heterotepala* & *P. icosandra*, are grouped together, and these two groups show polytomic relationships with *P. rugosa* and *P. thyrsiflora*.

It is interesting to note that *P. rugosa* and *P. sanguinea*, [subgenus *Pircuniopsis* Sect. *Pircuniopsis* of Nowicke (1969) and subgenus *Pircuniopsis* Sect. *Pircuniophorum* of Walter (1909)] are nested within the Clade III, while the other members, namely *P. dioica*, *P. tetramera* and *P. weberbaueri* [subgenus *Pircuniopsis* Sect. *Pseudolacca* of Walter (1909) and subgenus *Pircuniopsis* Sect. *Pircuniopsis* of Nowicke (1969)] form a separate Clade II with strong bootstrap support (96% BS).

Phytolacca acinosa [subgenus *Pircunia* Sect. *Pircuniastrum* of Walter (1909) and subgenus *Pircunia* Sect. *Pircunia* of Nowicke (1969)] and *P. dodecandra* [subgenus *Pircunia* Sect. *Pircunioides* of Walter (1909) and subgenus *Pircunia* Sect. *Pircunioides* of Nowicke (1969)] grouped together with *P. purpurascens* (54% BS), while *P. heptandra* [Subgen. *Pircunia* Sect. *Pircuniastrum* of Walter (1909) and subgenus *Pircunia* Sect. *Pircunia* of Nowicke (1969)] occupies basal most position in the MPTs as a separate clade.

In conclusion, this is the first inclusive study using molecular nrDNA ITS sequences to estimate phylogenetic relationships of *Phytolacca*. It is clearly evident that the phylogenetic trees resulting from the analysis of nrDNA ITS sequences are strongly supported as a monophyletic group (100% BS). However, the relationships among the species within the genus do not show harmony with the previous generic classification based on morphology. In the present analysis, a total number of 143 out of 654 (21%) sites of sequence data set were phylogenetically informative, so further sampling of additional taxon and addition of more regions are needed for the robust phylogeny of the genus *Phytolacca*. We herein based on the present analysis hypothesize that the intraspecific classification of *Phytolacca* should be recircumscribed into subgenus *Phytolacca* (*P. acinosa*, *P. americana*, *P. bogotensis*, *P. brachystachys*, *P. dodecandra*, *P. heterotepala*, *P. icosandra*, *P. meziana*, *P. octandra*, *P. purpurascens*, *P. rivinoides*, *P. rugosa*, *P. sanguinea* and *P. thyrsiflora*), subgenus *Pircuniopsis* (*P. dioica*, *P. tetramera* and *P. weberbaueri*), and *P. heptandra* should be treated under an independent subgenus. This treatment, as a hypothesis, however, needs testing and further data would help to clarify their true intraspecific affinities.

Acknowledgements

The first and second authors provided an equal contribution to this paper. Grant support (#2011-00402) from the Ministry of Education, Science and Technology, Government of South Korea to the second author is thankfully acknowledged. Authors thank to the curators of The University and Jepson Herbarium (UC), University of California, Berkeley, USA for providing plant materials for the study. The first and last authors acknowledge research supported by the King Saud University, Deanship of Scientific Research, College of Science, Research Center.

References

Ali, M.A, Al-Hemaid, F.M., Choudhary, R.K., Lee, J., Kim, S.Y. and Rub, M.A. 2013. Status of *Reseda pentagyna* Abdallah & A.G. Miller (Resedaceae) inferred from combined nuclear ribosomal and chloroplast sequence data. Bangladesh J. Plant Taxon. **20**(2): 233–238.

Ali, M.A., Gábor, G., Norbert, H., Balázs, K., Al-Hemaid, F.M.A., Pandey, A.K. and Lee, J. 2014. The changing epitome of species identification - DNA barcoding. Saudi J. Biol. Sci. **21**(3): 204–231.

Ali, M.A., Pandey, A.K., Al-Hemaid, F.M.A., Lee, J., Pandit, B., Kim, S.Y., Gyulai, G. and Rahman, M.O. 2015. Nuclear Sequences in Plant Phylogenetics. *In*: Ali, M.A., Gábor, G. and Al-Hemaid, F.M.A. (Eds), Plant DNA Barcoding and Phylogenetics. Lambert Academic Publishing, Germany, pp. 37–52.

APG III 2009. An update of the Angiosperm Phylogeny Group classification for the orders and families of flowering plants: APG III. Bot. J. Linn. Soc. **161**:105–121.

Brown, G.K. and Varadarajan, G.S. 1985. Studies in Caryophyllales I: Re-evaluation of classification of Phytolaccaceae *s.l.* Syst. Bot. **10**: 49–63.

Cuenoud, P., Savolainen, V., Chatrou, L.W., Powell, M., Grayer, R.J. and Chase, M.W. 2002. Molecular phylogenetics of Caryophyllales based on nuclear 18S rDNA and plastid *rbc*L, *atp*B, and *mat*K DNA sequences. Am. J. Bot. **89**: 132–144.

Downie, S., Katz-Downie, D. and Cho, K. 1997. Relationships in the Caryophyllales as suggested by phylogenetic analyses of partial chloroplast DNA ORF2280 homolog sequences. Am. J. Bot. **84**: 253–273.

Escalante, A.M., Santecchia, C.B., López, S.N., Gattuso, M.A., Gutiérrez, A., Delle Monache, F., González Sierra, M. and Zacchino, S.A. 2002. Isolation of antifungal saponins from *Phytolacca tetramera*, an Argentinean species in critic risk. J. Ethnopharmacol. **1**: 29–34.

Fassett, N. and Sauer, J. 1950. Studies of variation in the weed genus *Phytolacca*. 1. Hybridizing species in northeastern Colombia. Evolution **4**: 332–339.

Hall, T.A. 1999. BioEdit: a user-friendly biological sequence alignment editor and analysis program for Windows 95/98/NT. Nucleic Acids Symp. Ser. **41**: 95–98.

Heimerl, A. 1934. Phytolaccaceae. *In*: Engler, A. and Prantl, K. (Eds), Die naturlichen Pflanzenfamilien. 2nd edition. Leipzig, Wilhelm Engelmann, **4**: 135–164.

Hernández, M., Murace, M., Ringuelet, J., Petri, I., Gallo, D. and Arambarri, A. 2013. Effect of aqueous and alcohol extracts of *Phytolacca tetramera* (Phytolaccaceae) leaves on *Colletotrichum gloeosporioides* (Ascomycota). Bol. Soc. Argent. Bot. **48**(2): 201–209.

Lee, J., Kim, S.Y., Park, S.H. and Ali, M.A. 2013. Molecular phylogenetic relationships among members of the family Phytolaccaceae *sensu lato* inferred from internal transcribed spacer sequences of nuclear ribosomal DNA. Genet. Mol. Res. **12**(4): 4515–4525.

Lemma, A., Heyneman, D. and Kloos, H. (Eds). 1979. Studies on the molluscicidal and other properties of the endod plant, *Phytolacca dodecandra*, with special emphasis on the epidemiology of Schistosomiasis in Ethiopia and the possibility of localized control using endod as a molluscicide on a community self-help basis. Addis Ababa University, Ethiopia; University of California, San Francisco.

Nowicke, J.W. 1969. Palynotaxonomic study of the Phytolaccaceae. Ann. Miss. Bot. Gard. **55**: 94–363.

Santecchia, C., Escalante, A., Gattuso, M., Zacchino, M., Gutierrez Ravelo, A., Delle Monache, F. and Gonzalez Sierra, F. 2002. *Phytolacca tetrámera*, una fuente de saponinas triterpenoides. Revista Lat. Am. Quim. **28**: 246–247.

Sauer, J. 1951. Studies of variation in the weed genus *Phytolacca*. II. Latitudinally adapted variants within a North American species. Evolution **5**: 273–279.

Shu, S.L. 2003. *Phytolacca. In*: Wu, Z.Y., Raven, P.H. and Hong, D.Y. (Eds), Flora of China. Vol. **5**. Science Press, Beijing and Missouri Botanical Garden Press, St Louis, pp. 435–436.

Swofford, D.L. 2002. PAUP: Phylogenetic Analysis using Maximum Parsimony (and Other Methods). Version 4.0b 10. Sinauer, Sunderland, Massachusetts.

Thompson, J.D., Gibson, T.J., Plewniak, F., Jeanmougin, F. and Higgins, D.G. 1997. The Clustal X windows interface: Flexible strategies for multiple sequence alignment aided by quality analysis tools. Nucleic Acids Res. **24**: 4876–4882.

Umberto, Q. 2000. CRC World Dictionary of Plant Names: Common Names, Scientific Names, Eponyms, Synonyms, and Etymology. CRC press Taylor & Francis, New York, pp. 1–2065.

Walter, H. 1909. Phytolaccaceae. *In*: Enler, A. Pflanzenr. IV **83** (Heft 39): 1–154.

White, T.J., Bruns, T., Lee, S. and Taylor, J. 1990. Amplification and direct sequencing of fungal ribosomal RNA genes for phylogenetics. *In*: Innis, M.A., Gelfand, D.H., Sninksky, J.J. and White, T.J. (Eds), PCR Protocols: A Guide to Method and Amplifications. Academic Press, San Diego, California, pp. 315–322.

10

RELATIONSHIPS OF *ASTRAGALUS* L. IN SECTION SESAMEI BASED ON MORPHOLOGICAL CRITERIA AND MOLECULAR MARKERS

SHERIF M. SHARAWY[1,2] AND ABDELFATTAH BADR[3]

Botany Department, Faculty of Science, Ain Shams University, Cairo, Egypt

Keywords: *Astragalus*; Fabaceae; ISSR; RAPD; Section Sesamei.

Abstract

The relationships among five species and two varieties of *Astragalus* L. in the section Sesamei (Fabaceae) from Egypt and Saudi Arabia have been reassessed based on morphological variation and molecular polymorphism as revealed by RAPD and ISSR fingerprinting. The analysis of morphological variation delimited the examined taxa into two groups; one comprising samples representing *A. sinaicus*, *A. asterias* and *A. schimperi*, and the other is comprised of two samples of *A. stella* and six samples representing *A. tribuloides*. The grouping of *A. asterias* and *A. schimperi* based on morphological criteria indicates affinities between them that were not reflected in their previous treatments. Both morphological criteria and molecular markers indicated considerable distance between the samples of *A. stella* and *A. tribuloides*. The multiform nature of *A. tribuloides* is confirmed as *A. tribuloides* var. *mareoticus* is clearly differentiated from the type *A. tribuloides* and *A. tribuloides* var. *minutus*.

Introduction

Astragalus L. is the largest and most diverse genus of all angiosperms with more than 2,500 species distributed in arid and temperate regions of the Northern Hemisphere and South America (Podlech, 2008). It is particularly abundant in south western (SW) and south central (SC) Asia, western North America and South America (Maassoumi, 1998). The centre of origin and diversity of the genus is the drier mountainous parts of SW and SC Asia and the Himalaya (Maassoumi, 1998; Wojciechowski, 2005). In Egypt, *Astragalus* is represented by 32 species (Boulos, 1999) and in Saudi Arabia by 25 species (Migahid, 1996). The species in both countries are distributed in different phytogeographical regions and are delimited in several sections.

In the first comprehensive classification of the genus *Astragalus* presented by Bunge (1868), the annual species were assigned to two subgenera, *Trimeniaeus* Bunge and *Pogonophace* Bunge based on glabrous and barbellate stigma, respectively. In that classification subgenus *Trimeniaeus* included most of the species while subgenus *Pogonophace* contained only seven species. In recent taxonomic treatments of the genus, all annual species of *Astragalus* in the Old World were classified under subgenus *Trimeniaeus*, which has been considered to be monophyletic (Taeb *et al.*, 2007). Podlech (2008) classified the annual species of *Astragalus* into 14 sections including the section Sesamei DC. The section Sesamei is represented by five species in Egypt and five species in Saudi Arabia (Migahid, 1996; Boulos, 1999).

The molecular approaches to the taxonomy of *Astragalus* have been useful in constructing phylogenetic clades that help understand the evolutionary relationships and diversification in the genus (Wojciechowski, 2005; Kazempour Osaloo, *et al.*, 2005). Wojciechowski *et al.* (1999) have shown that some of the species-rich sections are monophyletic but other works indicated that none of the subgenera and large sections of the genus are monophyletic (Kazempour Osaloo, *et al.*, 2005).

---
[1]Corresponding author. E-mail: sherifsharaawy@yahoo.com
[1]Current Address: Biology department, Faculty of Science, Hail University, Hail, Saudi Arabia
[3]Botany and Microbiology Department, Faculty of Science, Helwan University, Cairo, Egypt

Random amplified polymorphic DNA (RAPD) and Inter simple sequence repeat (ISSR) markers are used for detecting genetic variation and species relationships (Williams *et al.*, 1990; Zietkiewicz *et al.*, 1994). In the genus *Astragalus* L., RAPD and ISSR markers have been applied in recent studies at the intra- and inter-specific relationship. ISSRs were chosen to assess genetic differentiation among population of the endemic species *Astragalus oniciformis* Barneby in the upper Snake River Plain of central Idaho in the USA (Alexander *et al.*, 2004). Intra- and inter-specific relationships within the *Astragalus microcephalus* complex were studied using RAPD (Mehrina *et al.*, 2005). High levels of genetic diversity were observed in three morphological types of *Astragalus membranaceus* (Fisch.) Bge. var. *mongholicus* (Bge.) Hsiao as revealed by ISSR (Xie *et al.*, 2009). Comparative analysis of molecular diversity of *Astragalus adsurgens* germplasm from north China was made using RAPD and ISSR Markers (Huang *et al.*, 2009). Anand *et al.* (2010) used ISSR, RAPD and DAMD (Directed amplification of mini-satellite DNA) to address the relationships among four closely related species of the *Astragalus rhizanthus* complex (i.e. *A. rhizanthus*, *A. candolleanus*, *A. malacophyllus* and *A. pindreensis*) from different parts of the Indian Himalaya and proved that these markers are potential to distinguish the closely allied species and to analyze the genetic diversity within and between the species of *Astragalus*. The objective of the present study is to clarify the systematic status of some taxa of *Astragalus* section Sesamei growing in Egypt and Saudi Arabia based on RAPD and ISSR polymorphism in addition to morphological variations.

Materials and Methods

Plant materials and scoring of morphological traits

The materials used in this study include 14 samples representing seven taxa of *Astragalus* section Sesamei collected from different localities in Egypt and Saudi Arabia (Table 1). The plant specimens have been identified following Boulos (1999) and Migahid (1996). The specimens of the examined taxa are deposited at the Herbarium of Botany Department, Faculty of Science, Ain Shams University, Cairo, Egypt and at the Museum of Biology Department, Faculty of Science, Hail University, Hail, Saudi Arabia.

A total of 45 morphological characters were considered, which include 32 two-state characters and 13 multi-state characters. The measurements and description of these characters were scored from at least five plants of each taxon. The characters and their states for morphological analysis are appended in Table 2.

DNA extraction

For DNA extraction, seeds of bulked samples of each of the studied taxa were germinated at 20°C for 15 days. Young seedlings were collected on ice and DNA was extracted from fresh young leaves using the CTAB method following the protocol of Saghai-Maroof *et al.* (1984).

RAPD fingerprinting

RAPD fingerprinting was performed using 20 arbitrary 10-mer random primers (Operon Technologies, Inc., USA). However, only ten primers gave clearly defined fingerprinting which are shown in Table 3. PCR was carried out using a Biocycler TC-S thermal cycler from HVD, Austria. The PCR reactions were developed in a total volume of 50 μl with the following components: 5 μl of 10X reaction buffer (75 mM Tris HCl, pH 9.0, 50 mM KCl, 20 nM $(NH_4)_2SO_4$ and 0.001% bovine serum albumin), 2 μl of 25 mM of each primer, 1 μl of Taq DNA polymerase (1U/μl), and 2 μl template DNA. The volume was completed to 50 μl with deionizied diethylpyrocarbonate (DEPC) water. The following PCR program was used: an initial denaturation of DNA was carried out at 94°C for 1 min, followed by 40 cycles of annealing at

37°C for 1 min, extension at 72°C for 2 min and a final extension at 72°C for 7 min. The RAPD products were resolved in 1.4% agarose gel in TAE buffer (0.04 M Tris-acetate, 1 mM EDTA; pH=8) at 100 volt for 60 min. A molecular size marker ranging from 530 to 1950 bp was used to estimate the size of resolved RAPD products. The gels were stained in 0.2 μg/ml ethidium bromide and photographed using a gel documentation system (Gel Doc BioRad 2000). Each experiment was repeated twice and only stable bands were scored.

Table 1. List of *Astragalus* L. taxa of the section Sesamei examined along with their locality.

Sl. No.	Taxon	Locality
1.	*Astragalus asterias* Stev. *ex* Ledeb. 1	Burg El-Arab, Egypt
2.	*A. asterias* Stev. *ex* Ledeb. 2	Hail-Al Jouf road, Saudi Arabia
3.	*A. schimperi* Boiss. 1	Saint Catherine, South Sinai, Egypt
4.	*A. schimperi* Boiss. 2	Aja Mountain, Hail, Saudi Arabia
5.	*A. sinaicus* Boiss. 1	Wadi El Arish, North Sinai, Egypt
6.	*A. sinaicus* Boiss. 2	Aja Mountain, Hail, Saudi Arabia
7.	*A. stella* L. 1	Wadi El Arish, Sinai, Egypt
8.	*A. stella* L. 2	Al Madinah-Makkah road, Saudi Arabia
9.	*A. tribuloides* Del. 1	Alexandria-Matruh Road, Egypt
10.	*A. tribuloides* Del. 2	Hail- Al Madinah road, Saudi Arabia
11.	*A. tribuloides* var. *mareoticus* Sirj. 1	Alexandria-Matruh Road, Egypt
12.	*A. tribuloides* var. *mareoticus* Sirj. 2	Hema Faid region, Hail, Saudi Arabia
13.	*A. tribuloides* var. *minutus* Boiss. 1	Saint Catherine, South Sinai, Egypt
14.	*A. tribuloides* var. *minutus* Boiss. 2	Al Madinah-Makkah road, Saudi Arabia

ISSR fingerprinting

Eight ISSR primers manufactured by the UBC (University of British Columbia, Canada) were used in the present study; the sequences of these primers are listed in Table 3. The amplification of ISSR markers was performed according to Nagoka and Ogihara (1997). The reaction mixture consisted of 12.5 μl Hot Start Master Mixture, 2.0 μl of primer (10 mM), 1.0 μl of template DNA (50 mg/μl), and filled up to 25 μl by ddH₂O. Amplification was carried out in a HVD thermocycler programmed as follows: 40 cycles after an initial cycle for 5 min at 94°C and each cycle consisted of a denaturation at 94°C for 2 min, annealing at 36°C for 1 min, extension at 72°C for 1 min followed by a final extension at 72°C for 7 min. The ISSR products were resolved in 1.5% agarose gel in TAE buffer (0.04 M Tris-acetate buffer, pH=8) at 100 volt for 60 min. A 1 kb ladder was used as DNA molecular size standard. ISSR bands were visualized on UV-trans-illuminator and photographed using gel documentation system (Gel Doc-BioRad 2000). Each experiment was repeated twice and only stable bands were scored.

Data analyses

The relationship among the examined taxa was estimated based on differences among them in morphological traits as well as ISSR and RAPD fingerprinting separately and in combination. The morphological traits were given codes ranging between 0 and 3 depending on the variation in the average value for the measured traits (Table 2). The RAPD and ISSR bands were scored as '1' and '0' for presence or absence, respectively. In order to construct trees elucidating the relationships among the examined taxa, the coded data were analyzed using UPGMA (Sokal and Michener, 1958) and the Neighbor-joining (Saitou and Nei, 1987) methods based on a distance matrix. All analyses were performed with NTSYS-pc (Rohlf, 2000).

Table 2. Morphological characters and their state used in the numerical analysis.

No.	Characters	Characters states
1.	Habit	Erect herb (0), prostrate herb (1)
2.	Length (cm)	0 – 10 (0), 10.1 – 20 (1), > 20 (2)
3.	Stem hairness	Tomentose (0), canescent (1), appressed (2), villous (3)
4.	Colour of stem hairs	White (0), white and black (1)
5.	Stipule length (cm)	0.5 (0), 0.51 – 1 (1), > 1 (2)
6.	Stipule width (cm)	0.1 – 0.5 (0), > 0.5 (1)
7.	Adnation of stipules	Free (0), adnate (1)
8.	Shape of stipules	Ovate (0), lanceolate (1), triangle (2)
9.	Stipule apex	Acute (0), acuminate (1)
10.	Stipule hairs	White (0), white and black (1)
11.	Leaf length (cm)	1 – 10 (0), > 10 (1)
12.	Leaf width (cm)	0.1 – 1 (0), > 1 (1)
13.	Leaf rachis	Imparipinnate (0), paripinnate (1)
14.	Colour of leaf hairs	White (0), white and black (1)
15.	Leaflet length (cm)	< 0.5 (0), 0.51 – 1 (1)
16.	Leaflet width (cm)	0.1 – 0.5 (0), > 0.5 (1)
17.	Leaflet upper surface	Glabrous (0), hairy (1)
18.	Leaflet arrangement	Opposite (0), alternate (1)
19.	Leaflet shape	Ovate (0), elliptic (1), lanceolate (2)
20.	Leaflet apex	Obtuse (0), acute (1), notched (2)
21.	Number of leaflets	1 – 10 (0), 11 – 20 (1), > 20 (2)
22.	Inflorescence type	Raceme (0), capitate (1)
23.	Peduncle length (cm)	0.1 – 5.0 (0), > 5 (1)
24.	Inflorescence hairs	White (0), black and white (1)
25.	Flower colour	White (0), purple (1), violet (2)
26.	Flower length (cm)	0.1 – 1 (0), 1.1 – 1.5 (1), > 1.5 (2)
27.	Calyx length (cm)	< 0.5 (0), 0.51 – 1.0 (1)
28.	Colour of calyx hairs	White (0), white and black (1),
29.	Stamen length (cm)	0.1 – 0.5 (0), 0.5 – 1 (1), > 1 (2)
30.	Ovary length (cm)	0.1 – 0.5 (0), 0.5 – 1 (1), > 1 (2)
31.	Ovary width (cm)	0.1 (0), 0.2 (1)
32.	Pod length (cm)	0.1 – 2 (0), > 2 (1)
33.	Pod width (cm)	0.1 – 0.5 (0), > 0.5 (1)
34.	Pod pedicel	Absent (0), shorter than pod (1), longer than pod (2)
35.	Pod texture	Glabrous (0), hairy (1)
36.	Pod surface	Membranous (0), wrinkled (1)
37.	Pod dorsal suture	Obtuse (0), grooved (1), furrowed (2)
38.	Pod ventral suture	Obtuse (0), furrowed (1)
39.	Pod apex	Acute (0), beaked (1)
40.	Number of seeds	1–10 (0), > 10 (1)
41.	Seed length (cm)	0.1 – 0.2 (0), > 0.2 (1)
42.	Seed width (cm)	0.1 – 0.2 (0), > 0.2 (1)
43.	Seed shape	Reniform (0), quadrate (1)
44.	Seed colour	Yellow (0), brown (1)
45.	Seed surface	Smooth (0), Irregular (1)

Table 3. RAPD and ISSR primers used for DNA fingerprinting in *Astragalus* L. taxa.

	RAPD primers			ISSR primers	
No.	Primer code	Primer base sequence	No.	Primer code	Primer base sequence
1	A14	5′TCT GTG CTGG 3′	1	UBC808	$(AG)_8C$
2	B17	5′AGG GAA CGAG 3′	2	UBC809	$(AG)_8G$
3	OPA01	5′CAG GCC CTTC 3′	3	UBC810	$(GA)_8T$
4	OPB07	5′GCT GAC GCAG 3′	4	UBC812	$(GA)_8A$
5	OPB20	5′GGA CCC TTAC 3′	5	UBC 830	$(TG)_8G$
6	F01	5′ACG GAT CCTG 3′	6	UBC840	$(GA)_8CT$
7	O04	5′AAG TCC GCTC 3′	7	UBC848	$(CA)_8AG$
8	O06	5′CCA CGG GAAG 3′	8	UBC855	$(AC)_8CT$
9	O08	5′CCT CCA GTGT 3′			
10	O16	5′TCG GCG GTTC 3′			

Results and Discussion

RAPD and ISSR fingerprinting analyses

A total of 91 RAPD bands were generated by 10 primers in 14 samples of *Astragalus* taxa investigated. Of these 68 bands are polymorphic and 23 are monomorphic. The polymorphic bands include 12 unique bands that have been revealed by seven primers (Table 4). The highest number of both total bands (20) and polymorphic bands (17) was produced by the primer OPB07. The RAPD fingerprints generated by the primer OPB07 is shown in Fig. 1. The primer A14, on the other hand, produced the highest number of monomorphic and unique bands (Table 4). The least number of bands (4 bands) was generated by two primers, namely O04 and O08; the number of polymorphic bands was 2 for the primer O04 and only 1 for the primer O08 with 50% and 25% polymorphism respectively (Table 4).

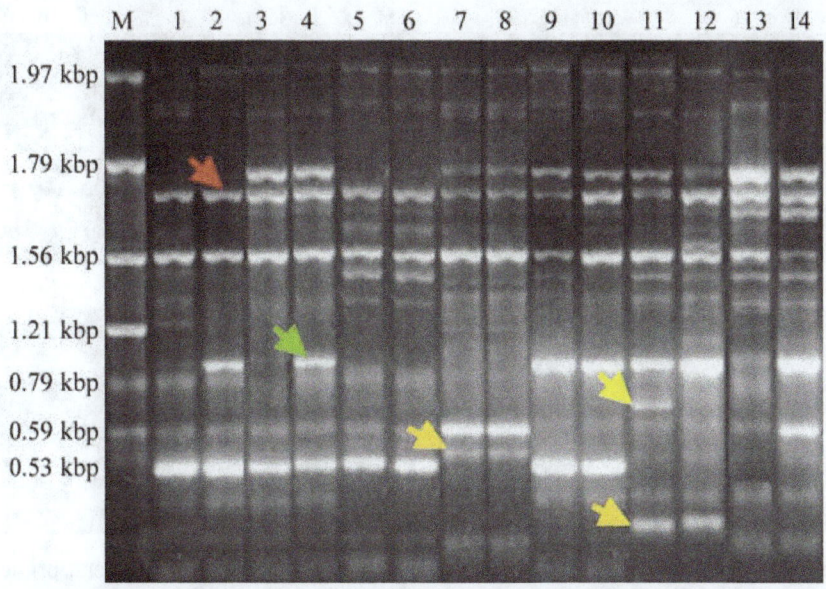

Fig. 1. RAPD fingerprints of the studied 14 samples of *Astragalus* L. as revealed by the primer OPB07. The lane to the left is a molecular size marker. Numbers on lanes 1-14 correspond to the serial numbers of samples as numbered in Table 1. First arrow indicates a monomorphic band, second arrow indicates polymorphic band, other arrows indicate unique band.

The number of amplified bands generated by RAPD markers and their molecular size are given in Table 5. The primer A14 generated the highest number of bands ranging from 11 in *A. stella* and the varieties of *A. tribuloides* to 13 in *A. schimperi* and *A. sinaicus*. OPB07 generated a total of 121 bands ranging from 6 in *A. asterias* to 12 in *A. tribuloides* var. *minutes*. In contrast, the least number of bands were produced by the primer O06 (Table 5).

Table 4. Number and types of amplified RAPD bands generated in the examined 14 samples of *Astragalus* L.

Types of bands	RAPD Primers and number of bands										
	A14	B17	OPA01	OPB07	OPB20	F01	O04	O06	O08	O16	Total
Monomorphic	4	2	3	2	2	3	2	1	3	1	23
Unique	4	1	1	1	2	0	0	1	0	1	11
Polymorphic	10	4	5	17	9	4	2	3	1	4	59
Total bands	18	7	9	20	13	7	4	5	4	6	93
% of polymorphism	77.8	71.4	66.7	90	84.6	57.1	50	80	25	83.3	75.3

Eight ISSR primers produced a total of 37 bands including only 14 polymorphic bands (Table 6; Fig. 2). The number of bands ranged from 3 as revealed by the three primers 809, 848 and 855 to 7 revealed by the primer 810; all of the bands produced by the two primers 809 and 812 were monomorphic. The primer 830 (Fig. 2C) produced a band that in all taxa except the two samples

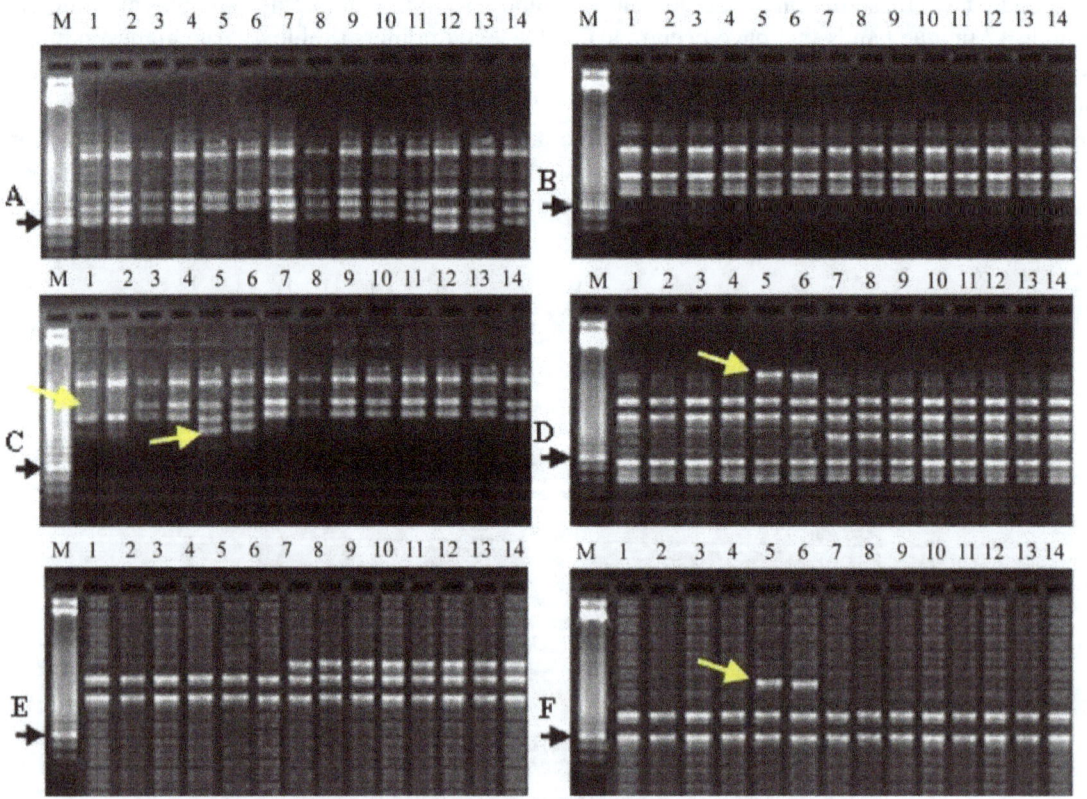

Fig. 2. ISSR fingerprints for 14 samples of *Astragalus* L. as revealed by six ISSR primers; primer codes are as follows: A = Primer UBC810, B = Primer UBC812, C = Primer UBC830, D = Primer UBC840, E = Primer UBC848, F = Primer UBC855 (see Table 4). Short arrows to the lane M indicate 250 bp and long arrows shows bands unique to one species. Number on lanes 1-14 correspond to the serial numbers of *Astragalus* taxa as numbered in Table 1.

Table 5. Number of amplified bands and their molecular size (in bp) produced in the 14 samples of *Astragalus* taxa as revealed by RAPD primers.

No.	Taxon	RAPD Primers, number and size range of bands										
		A14	B17	OPA01	OPB07	OPB20	F01	O04	O06	O08	O16	Total Bands
1.	*Astragalus asterias* 1	12 (530-1900)	2 (700-1700)	6 (500-1500)	6 (530-1970)	8 (400-1300)	5 (500-900)	2 (750-900)	2 (550-750)	4 (590-1200)	4 (1400-1700)	51
2.	*A. asterias* 2	12 (530-1900)	2 (700-1700)	5 (500-1500)	6 (530-1970)	8 (400-1300)	5 (500-900)	2 (750-900)	2 (550-750)	4 (590-1200)	4 (1400-1700)	50
3.	*A. schimperi* 1	13 (530-1900)	2 (700-1700)	5 (500-1500)	8 (530-1970)	7 (350-1250)	5 (500-900)	2 (750-900)	2 (550-750)	4 (590-1200)	4 (1400-1700)	52
4.	*A. schimperi* 2	13 (530-1900)	2 (700-1700)	5 (500-1500)	7 (530-1970)	7 (350-1250)	5 (500-900)	2 (750-900)	2 (550-750)	4 (590-1200)	4 (1400-1700)	52
5.	*A. sinaicus* 1	13 (530-1900)	3 (700-1700)	5 (500-1500)	9 (530-1970)	7 (350-1250)	4 (500-900)	2 (750-900)	2 (550-750)	4 (590-1200)	5 (1400-1700)	54
6.	*A. sinaicus* 2	12 (530-1900)	4 (700-1700)	5 (500-1500)	9 (530-1970)	7 (350-1250)	4 (500-900)	2 (750-900)	3 (550-750)	4 (590-1200)	4 (1400-1700)	54
7.	*A. stella* 1	11 (530-1900)	3 (700-1700)	7 (500-1900)	9 (450-1970)	9 (350-1700)	5 (500-900)	4 (750-900)	3 (500-750)	4 (590-1200)	3 (1350-1700)	58
8.	*A. stella* 2	11 (530-1900)	3 (700-1700)	7 (500-1900)	9 (450-1970)	9 (350-1700)	5 (500-900)	4 (750-900)	3 (500-750)	4 (590-1200)	3 (1350-1700)	58
9.	*A. tribuloides* 1	13 (530-1900)	3 (700-1700)	4 (500-1900)	8 (550-1970)	10 (450-1700)	5 (500-900)	4 (750-900)	2 (500-750)	4 (590-1200)	3 (1350-1700)	56
10.	*A. tribuloides* 2	13 (530-1900)	3 (700-1700)	4 (500-1900)	8 (550-1970)	9 (450-1700)	5 (500-900)	4 (750-900)	3 (500-750)	4 (590-1200)	3 (1350-1700)	56
11.	*A. tribuloides* var. *mareoticus* 1	12 (530-1900)	3 (700-1700)	4 (500-1900)	9 (530-1970)	9 (450-1700)	5 (500-900)	4 (750-900)	3 (500-750)	4 (590-1200)	3 (1350-1700))	56
12.	*A. tribuloides* var. *mareoticus* 2	11 (530-1900)	3 (700-1700)	5 (500-1900)	9 (530-1970)	10 (450-1700)	5 (500-900)	4 (750-900)	3 (500-750)	4 (590-1200)	3 (1350-1700)	57
13.	*A. tribuloides* var. *minutus* 1	11 (530-1900)	3 (700-1700)	5 (500-1900)	12 (480-1970)	9 (450-1700)	5 (500-900)	4 (750-900)	3 (500-750)	4 (590-1200)	3 (1350-1700)	59
14.	*A. tribuloides* var. *minutus* 2	11 (530-1900)	3 (700-1700)	5 (500-1900)	12 (480-1970)	9 (450-1700)	5 (500-900)	4 (750-900)	3 (500-750)	4 (590-1200)	3 (1350-1700)	59

of *A. asterias* (lanes 1&2); the same primer, produced a band in the profile *A. sinaicus* (lanes 5 & 6) that were absent in the profile of other taxa. The two samples of the same species are also clearly distinguished by two bands in profile of primer 840 (Fig. 2D). In the profile of primer 848 (Fig. 2E), one band was evident in the ISSR profile of the two samples of *A. stella* (lanes 7-8) and the six samples representing the three varieties of *A. tribuloides* (lanes 9-14) and was absent from the profile of the taxa representing *A. asterias*, *A. schimperi* and *A. sinaicus* (lanes 1-6). In the profile of primer 855 (Fig. 2F), it is apparent that the ISSR profiling clearly differentiated *A. sinaicus* (lanes 5 & 6) by the presence of two bands that are absent in all other taxa. A glimpse on the ISSR profiling in all samples indicates that *A. asterias* (lanes 1-2) is characterized by the absence of one band in the profile of primer 830 (Fig. 2C) and *A. sinaicus* (lanes 5-6) is distinguished by presence of three unique bands in profile of primers 830, 840 and 855.

Table 6. Number and type of amplified bands generated by the eight primers in *Astragalus* section Sesamei.

Types of bands	ISSR Primers and number of bands								Total
	808	809	810	812	830	840	848	855	
Monomorphic	4	3	2	4	2	4	2	2	23
Unique	0	0	0	0	0	0	0	0	0
Polymorphic	1	0	5	0	4	2	1	1	14
Total bands	5	3	7	4	6	6	3	3	37
% of polymorphism	20	0	71.4	0	66.7	33.3	33.3	33.3	37.8

Relationship among Astragalus taxa based on morphological variation:

The 14 samples of *Astragalus* are clearly divided into two groups in the UPGMA tree (Fig. 3), one comprising the taxa of *A. sinaicus*, *A. asterias* and *A. schimperi* and the other is comprised of taxa representing *A. stella* and the six samples representing *A. tribuloides* and its two varieties *A. tribuloides* var. *mareoticus* and *A. tribuloides* var. *minutus*. In the former group, the two samples of *A. sinaicus* are clearly delimited from the four samples representing *A. asterias* and *A. schimperi*. In the other group, the two samples representing *A. stella* are delimited from the other six samples representing *A. tribuloides*, *A. tribuloides* var. *mareoticus* and *A. tribuloides* var. *minutus*. The level of distance that separates the taxa of *A. tribuloides* exceeds the levels that separate the taxa representing *A. asterias* and *A. schimperi* (Fig. 3).

Relationship among Astragalus taxa based on RAPD and ISSR polymorphism:

The analyses of RAPD and ISSR data show that the two samples representing *A. sinaicus* are clearly delimited from the other taxa (Fig. 4). The other 12 samples are divided into two subgroups; one comprised of four samples representing the two species *A. asterias* and *A. schimperi*. The second subgroup includes the two samples representing *A. stella* and the six samples representing *A. tribuloides*. In this subgroup the two samples of the former species are clearly separated from the six samples representing *A. tribuloides* at the distance of 4.80. The two samples representing *A. tribuloides* var. *mareoticus* are separated from the four samples representing *A. tribuloides* and *A. tribuloides* var. *minutus* at a distance of 3.70. The separation of the two samples representing *A. sinaicus* is clearly associated with the presence of three ISSR bands that are confined to material of this species and absent in the other taxa (Fig. 2C, D & F).

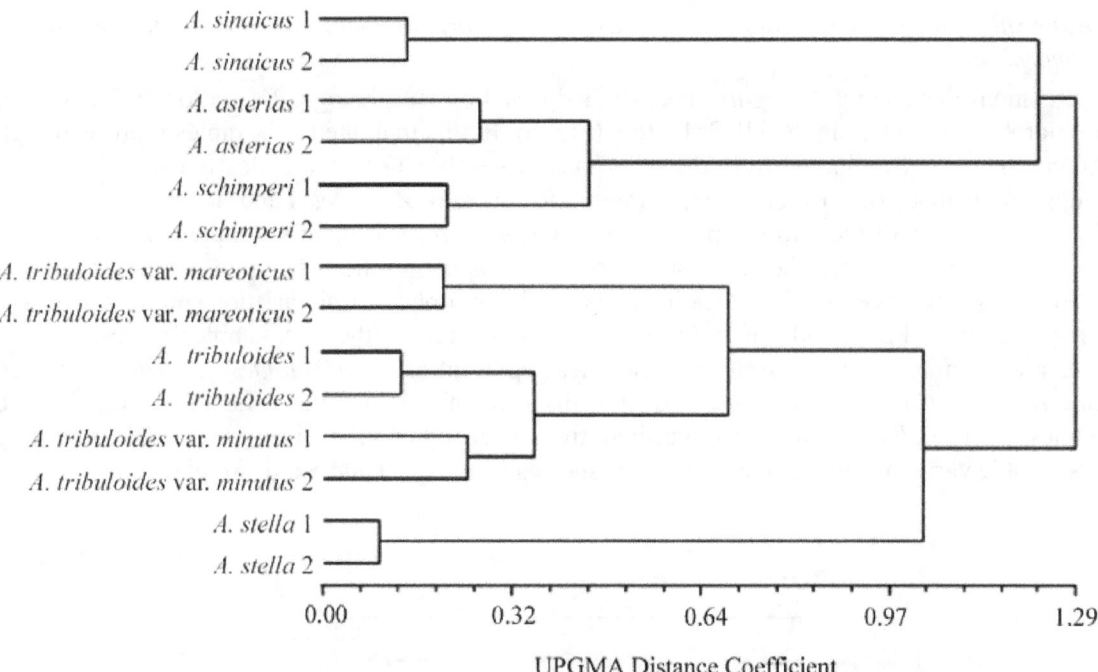

Fig. 3. UPGMA tree illustrating the relationships among *Astragalus* taxa based on morphological characters.

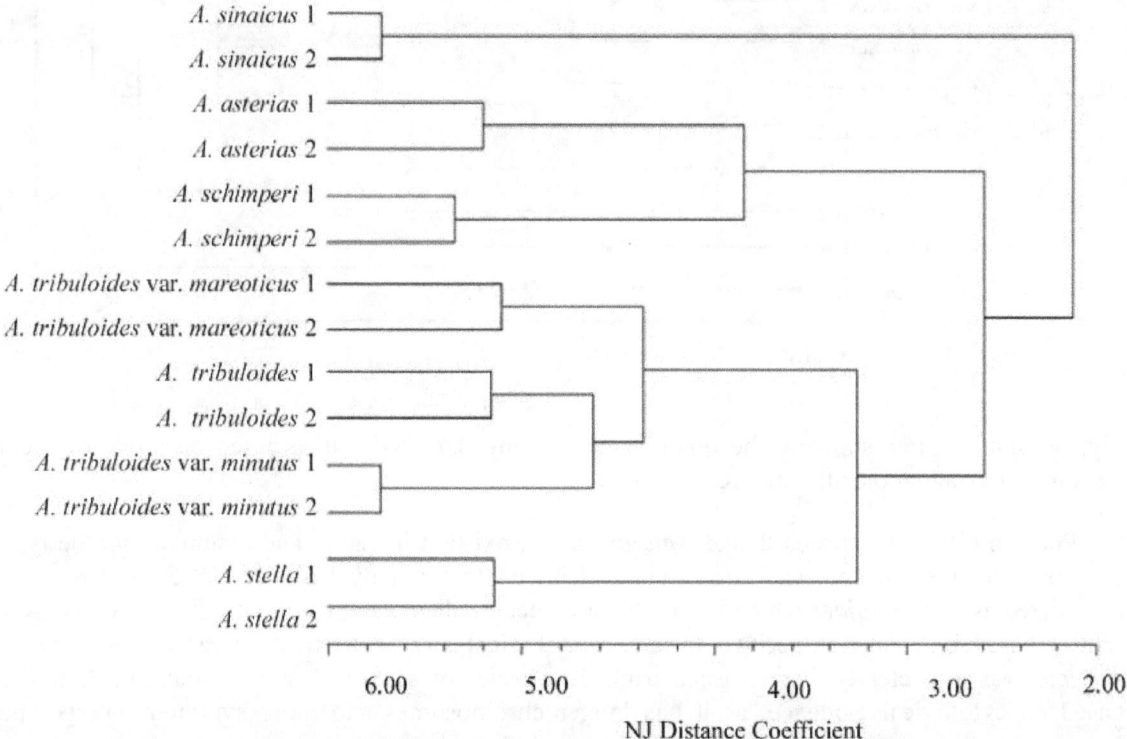

Fig. 4. Neighbour joining tree illustrating the relationships among *Astragalus* taxa based on RAPD and ISSR markers.

Relationship among Astragalus taxa based on morphological variation and molecular polymorphism:

Relationships among *Astragalus* taxa studied based on morphological variation and molecular polymorphism is shown in the UPGMA tree (Fig. 5). In this tree, the two samples representing *A. sinaicus* are clearly delimited from the other taxa. The other 12 taxa are clearly divided into two groups at a distance of 1.15, one comprising the four taxa of *A. asterias* and *A. schimperi* and the other is comprised of the two samples representing *A. stella* and the six samples representing *A. tribuloides*. It is noted that the two samples of *A. schimperi*, in the first group, are delimited at a relatively high distance of 0.81 indicating considerable morphological variation among material of this species from Egypt and Saudi Arabia. In the other group, the two samples representing *A. stella* are delimited from the other six samples representing *A. tribuloides, A. tribuloides* var. *mareoticus, A. tribuloides* var. *minutus* at a distance of 1.15 on the distance scale. The two samples of *A. stella* are also distinguished from each other at a distance of 0.70 indicating considerable variation among material of this species from Egypt and Saudi Arabia.

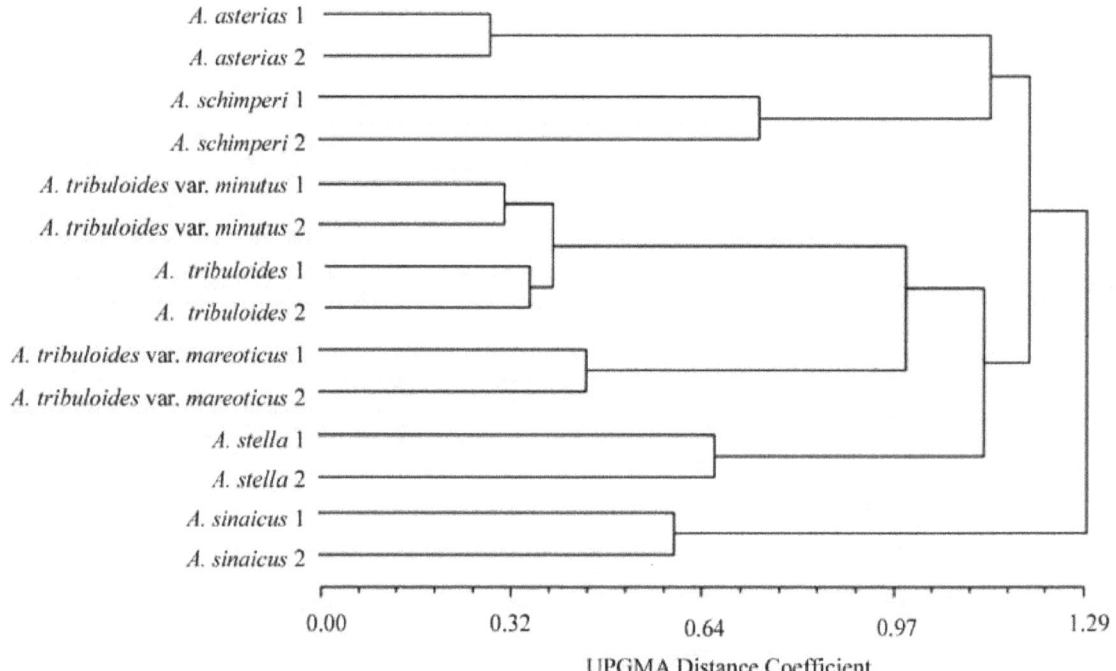

Fig. 5. UPGMA tree showing the relationships among *Astragalus* taxa based on morphological characters and molecular markers.

Podlech (1991) suggested that *A. sinaicus* is not existent in Egypt and assumed that the type may be from Greece and was erroneously attributed to Sinai by Boissier (1872) and may be considered as *A. tribuloides* which is a multiform species. This view was contradicted by Sharawy (2001) based on evidence derived from morphological and anatomical characters. *Astragalus sinaicus* was also clearly distinguished from the species of section Sesamei including *A. stella*, based on cytological evidence, as it has longer chromosomes and more symmetric karyotype compared to the other species (Badr and Sharawy, 2007). The analysis of morphological variation in the present study delimited *A. sinaicus* in a major group that also includes *A. asterias* and *A. schimperi* but remained distinguished as a separate identity. The distinction of this species is also clearly reflected in the relationship based of the analysis of ISSR and RAPD fingerprinting polymorphism.

The grouping of *A. asterias* and *A. schimperi* based on morphological variation and molecular polymorphism is congruent with their grouping together based on the analysis of seed protein electrophoretic profile (Al-Nowaihi *et al.*, 2002). However, evidence from seed protein electrophoretic analysis also indicated the grouping of *A. asterias* with *A. tribuloides* (Al-Nowaihi *et al.*, 2002) that is not supported by molecular evidences expressed by the analysis of morphological variation and molecular polymorphism which is correlated with similarities between these two species in spermoderm characteristics (Sharawy, 2001). *A. asterias* possesses sessile leaves and fruits with double indumentum (Sharawy *et al.,* 2003) that distinguish it from other species in section Sesamei, which can also be distinguished based on pollen characters (Saad and Taia, 1988).

Gazer (1993) divided the species of section Sesamei into four groups; i.e. *Astragalus asterias* group, *A. schimperi* group, *A. sinaicus* group and *A. stella* group; the latter group also comprised *A. tribuloides*. In the present investigation, the recognition of *A. sinaicus* and *A. stella* as distinct groups is supported by the relationships as expressed in the UPGMA trees based on morphological and molecular evidences. Both morphological criteria and molecular markers indicated considerable distance between the two samples of *A. stella* and the six samples of *A. tribuloides* and its two varieties, i.e. *A. tribuloides* var. *mareoticus* and *A. tribuloides* var. *minutus*. The distance levels among these varieties confirm the observations by Boissier (1872) and Podlech (1986) that *A. tribuloides* is a multiform species. In the present investigation *A. tribuloides* var. *mareoticus* is clearly distinct from *A. tribuloides* and *A. tribuloides* var. *minutus*.

References

Alexander, J.A., Liston, A. and Popovich, S.J. 2004. Genetic diversity of the narrow endemic *Astragalus oniciformis* (Fabaceae). Am. J. Bot. **91**: 2004-2012.

Al-Nowaihi, A.S., Khalifa, S.F., Badr, A. and Sharawy, S.M. 2002. Species relationships of *Astragalus* L. in Egypt based on storage seed protein electrophoretic criteria. Proc. 2nd. Int. Conf. Biol. Sci. Faculty of Science, Tanta University **2**: 174-188.

Anand, K.K., Srivastava, R.K., Chaudhary, L.B. and Singh, AK.I. 2010. Delimitation of species of the *Astragalus rhizanthus* complex (Fabaceae) using molecular markers RAPD, ISSR and DAMD. Taiwania **55**: 197-207.

Badr, A. and Sharawy, S.M. 2007. Karyotype analysis and systematic relationships in the Egyptian *Astragalus* L. (Fabaceae). International J. Bot. **3**: 147-159.

Boissier, E. 1872. Flora Orientals, Vol. **2**. Genevae and Basiliae, Lugduni.

Boulos, L. 1999. Flora of Egypt **1**: 320-336. Al-Hadara Publishing, Cairo, Egypt.

Bunge, A. 1868. Generis Astragali species Gerontogeae, Pars prior. Claves diagnosticae. Mém. Acad. Imp. Sci. St. Pétersb. **11**: 1-140.

Gazer, M. 1993. Revision of *Astragalus* L. Sect. Sesamei DC. (Leguminosae). Sendtnera **1**: 69-155.

Huang, L., Chen, Z. and Zhang, X. 2009. A comparative analysis of molecular diversity of erect milkvetch (*Astragalus adsurgens*) germplasm from north China using RAPD and ISSR markers. Biochem. Genet. **47**: 92-99.

Kazempour Osaloo, S., Maassoumi, A.A. and Murakami, N. 2005. Molecular systematics of the Old World *Astragalus* (Fabaceae) as inferred from nrDNA ITS sequence data. Brittonia **57**: 367-381.

Maassoumi, A.A. 1998. *Astragalus* L. in the World. Research Institute of Forests and Rangelands, Tehran, Iran.

Mehrnia, M., Zarre, S. and Sokhan-Sanj, A. 2005. Intra- and interspecific relationships within the *Astragalus microcephalus* complex (Fabaceae) using RAPD. Biochem. Syst. Ecol. **33**: 149-158.

Migahid, A.M. 1996. Flora of Saudi Arabia. 4th ed. King Saud University, Riyadh.

Nagoka, T. and Ogihara, Y. 1997. Applicability of inter-simple sequence repeat polymorphisms in wheat for use as DNA markers in comparison to RFLP and RAPD markers. Theor. Appl. Genet. **94**: 597-602.

Podlech, D. 1986. Taxonomic and phytogeographical problems in *Astragalus* of the Old World and southwest Asia. Proc. Royal Soc. Edinburgh **89**: 37-43.

Podlech, D. 1991. The systematics of annual species of the genus *Astragalus* L. (Leguminosae). Flora *et* Vegetatio Mundi **9**: 1-18.

Podlech, D. 2008. The genus *Astragalus* L. (Fabaceae) in Europe with exclusion of the former Soviet Union. Feddes Repertorium **119**: 310-387.

Rohlf, F.J. 2000. NTSYS-pc: Numerical Taxonomy System. Ver. 2.1. Exeter Publishing, Ltd. Setauket, New York.

Saad, S.I. and Taia, W.K. 1988. Palynological studies of some species in the genus *Astragalus* L. Leguminosae in Egypt. Arab. Gulf. J. Sci. Res. **B6**: 227-243.

Saghai-Maroof, M.A., Soliman, K.M., Jorgensen, R.A. and Allard, R.W. 1984. Ribosomal DNA spacer-length polymorphisms in barly mendelian inheritance, chromosomal location and population dynamics. Proc. Natl. Acad. Sci. USA **81**: 8014-8018.

Saitou, N. and Nei, M. 1987. The neighbor-joining method: a new method for reconstruction phylogenetic trees. Mol. Biol. Evol. **4**: 40-42.

Sharawy, S.M. 2001. Taxonomic studies on interspecific and infraspecific relationships in the genus *Astragalus* in Egypt. Ph.D. Thesis, Ain Shams University, Cairo, Egypt.

Sharawy, S.M., Mourad, M.M. and Al-Nowaihi, A.S. 2003. The assessment of the morpho-anatomical characters of the spermoderm in delimitation of some *Astragalus* taxa growing in Egypt. Bull. Fac. Sci. Assiut. Univ. **32**: 325-346.

Sokal, R.R. and Michener, C.D. 1958. A statistical method for evaluating systematic relationships. Univ. Kansas Sci. Bull. **28**: 1409-1438.

Taeb, F., Zarre, S., Podlech, D., Tillich, H., Kazempour Osaloo, S. and Maassoumi, A. 2007. A contribution to the phylogeny of annual species of *Astragalus* (Fabaceae) in the Old World using hair micromorphology and other morphological characters. Feddes Repertorium **118**: 206-227.

Williams, J.G.K., Kubelik, A.R., Livak, K.J., Rafalski, J.A. and Tingey, S.V. 1990. DNA polymorphisms amplified by arbitrary primers are useful as genetic markers. Nucleic Acids Research **18**: 6531-6535.

Wojciechowski, M.F., Sanderson, M.J. and Hu, J.M. 1999. Evidence on the monophyly of *Astragalus* (Fabaceae) and its major subgroups based on nuclear ribosomal DNA ITS and chloroplast DNA *trn*L intron data. Systematic Botany **24**: 409-437.

Wojciechowski, M.F. 2005. *Astragalus* (Fabaceae): a molecular phylogenetic perspective. Brittonia **57**: 382-396.

Xie, X., Hu, Y., Wang, L., Yang, J., Li, Y. and Peng, M. 2009. Genetic diversity in three morphological types of *Astragalus membranaceus* (Fisch.) Bge. var. *mongholicus* (Bge.) Hsiao as revealed by inter simple sequence repeat markers. African J. Biotech. **8**: 4490-4494.

Zietkiewicz, E., Rafalski, A. and Labuda, D. 1994. Genome fingerprinting by simple sequence repeat (SSR) - anchored polymerase chain reaction amplification. Genetics **20**: 176-183.

11

CONFIRMING THE IDENTITY OF NEWLY RECORDED *NYMPHAEA RUBRA* ROXB. *EX* ANDREWS DISCERNING FROM *NYMPHAEA PUBESCENS* WILLD. USING MORPHOMETRICS AND MOLECULAR SEQUENCE ANALYSES

D.P.G. Shashika K. Guruge[1], Deepthi Yakandawala[2] and Kapila Yakandawala[3]

Department of Botany, Faculty of Science, University of Peradeniya, Sri Lanka

*Keywords: mat*K; Morphometric Analysis; Nymphaeaceae; *psb*A-*trn*H; Water-lilies; Sri Lanka.

Abstract

A multivariate statistical analysis was carried out to evaluate the morphological variation between *Nymphaea pubescens* Willd., and a deep purplish red flowered *Nymphaea* that occur in Sri Lanka. The plant resembles *N. rubra* Roxb. ex Andrews, a species that had been sometimes circumscribed as a variety under *N. pubescens* Willd. DNA sequences data of *matK* and *psbA-trnH* regions were used to obtain further support. Morphological data were scored from collected samples and analyzed using PAST software. Extracted DNA were amplified for *matK* and *psbA-trnH* gene regions. Obtained sequences were matched with the related accessions deposited in the GenBank. Multivariate analysis supported the recognition of deep purplish red flowered *Nymphaea* as a different species from *N. pubescens*, and was identified as *N. rubra* based on literature. GenBank accessions for the *matK* region of *N. rubra* showed 99% similarity while it gave only a 96% similarity for *N. pubescens* with query coverage of 97% and 96% respectively, corroborating with the morphological analysis. Comparison of the sequence divergence between *N. pubescens* and *N. rubra* sequences indicated a 95% similarity for *matK* gene region while 92% similarity for *psbA-trnH* gene region. The sequences generated during the present study would provide additional reference sequences for the two taxa.

Introduction

The genus *Nymphaea* L. (*Nymphaeaceae* Salisb.) or Water-lilies comprise of about 40-50 species and is widespread in tropical and temperate regions covering vast extents of natural water-bodies. All are aquatics with perennial or annual rhizomes (Jaime *et al.,* 2000). Species of *Nymphaea* show a high morphological plasticity where the size of leaves and flowers are thought to be strongly dependent on hydrological and edaphic conditions (Polina and Alexy, 2007). They grow in open waters of large swamps, lakes, ponds, shallow ditches, and also in marshes. The species of *Nymphaea* may be either day- blooming or night-blooming. The flowers are showy and born solitarily, containing numerous petals, stamens, and many carpels. The genus *Nymphaea* is a taxonomically difficult group; many species are believed to have numerous subspecies, chromosomal races & forms of hybrids and of artificial origin (Polina and Alexy, 2007). The plants are very popular as ornamental aquatics in the landscape industry.

[1]Postgraduate Institute of Science, University of Peradeniya, Sri Lanka
[2]Corresponding author. Email: deepthiyakandawala@gmail.com
[3]Department of Horticulture & Landscape Gardening, Faculty of Agriculture & Plantation Management, Wayamba University of Sri Lanka.

Among the diverse members of the family, *N. nouchali* Burm.f., *N. rubra* Roxb. ex Andrews, *N. pubescens* Willd, and *N. alba* L. are some of the most widely spread species in Asia (La-ongsri *et al*, 2009). *Nymphaea rubra* is common throughout the temperate and tropical Asia, such as in Bangladesh, India, Taiwan and Thailand, especially in shallow lakes and ponds. The species have somewhat big flowers compared to many other *Nymphaea* species, and it prefers to grow in non-acidic water above 15°C (Hossain *et al*., 2007).According to the Revised Handbook to the Flora of Ceylon, the family Nymphaeaceae is represented in Sri Lanka only by the genus *Nymphaea*, with two species, *N. nouchali* Burm. f., and *N. pubescens* Willd. (Dassanayake, 1996). Other than these native species, during field visits, a deep purplish red flowered *Nymphaea* species with a morphological resemblance to *N. pubescens* was also encountered in natural water bodies in the dry lowland of the country. Many morphological features that are described under the *N. pubescens* (Dassanayake, 1996) overlap with this *Nymphaea* species. The plant has been referred to as *N. pubescens* variety *rubra* by de Vlas and de Vlas-de Jong in 2008. According to The Revised Handbook to the Flora of Ceylon (Dassanayake, 1996) the flower petals of *N. pubescens* are white, purplish pink or red, where inner petals are smaller. Leaf upper surface is glossy dark green and dark purplish green, velvety on leaf lower surface with very prominent veins. The petiole is red-brown, while the pedicel bears short prickles. However the presence of short prickles on the pedicels is a mis-conception as *N. pubescens* never poses prickles but *Nelumbo nucifera,* a species belonging to the family Nelumbonaceae, instead. The filament colour is described as yellowish white becoming deeper yellow distally, or pale purplish pink to crimson. Although the above description is accommodating many characters of *N. pubescens*, the description seems to include some characters of those of the deep purplish red flower species of *Nymphaea* as well. Characters such as red- brown petiole, dark purplish green lower surface with velvety appearance and highly prominent veins are more towards the plants with deep purplish red flowered *Nymphaea* rather than *N. pubescens*. On the other hand, according to literature, this deep purplish red flowered *Nymphaea* species share morphological similarities with *N. rubra*, a species that had not been recognized as occurring in the island during the revision of the Flora. According to Conard (1905), *N. rubra* possess deep purplish red coloured flowers with cinnabar red stamens, and the reddish leaves becoming greenish with the age, and rarely producing fruits or seeds. Mitra and Subramanyam (1982), questioned the treatment of *N. rubra* as a true species at par with other sexually reproducing species because of its failure to set fruits/seeds in nature. According to Gupta (1980), *N. rubra* has two cytotypes, one which is highly fertile and another nearly sterile. Further, La-ongsri *et al.* (2009), describes *N. rubra* as, leaf dark reddish above and below, nine pairs of prominent and angular veins below, petiole green or reddish-brown, and a deep purplish red flower bearing orange or cinnabar-red stamens, becoming brownish with age. *Nymphaea pubescens* and *N. rubra* are two closely related taxa (Jeremy *et al.*, 2010).

Hence, a multivariate statistical analysis was carried out to evaluate the morphological variation between the ambiguous taxa and described *N. pubescens*, and further DNA sequences data of *matK* and *psbA-trnH* regions were used in verification of the identity between *N. pubescens* and the deep purplish red flowered *Nymphaea* species occurring in Sri Lanka. The *matK* is one of the rapidly evolving coding region in the plastid genome, while Chloroplast non-coding intergenic *psbA-trnH* spacer has recently become a popular tool in plant molecular phylogenetic studies at low taxonomic levels (Biswal *et al*., 2012).

Materials and Methods
Sample collection

Live plant material of the two *Nymphaea* species, including populations with both white and pink flowered *N. pubescens*, and deep purplish red flowered *Nymphaea* species, were collected from 50 different locations covering all the three major climatic zones of the island. The map showing the field localities are given in Fig. 1. From each locality, a minimal of five specimens were collected. All the collected populations were treated separately with a different acronym; DPRN (Deep Purplish Red flowered *Nymphaea* species), NPW (*N. pubescens* White) and NPR (*N. pubescens* Pink), for easy references. The collected specimens were examined in detail in the laboratory for different morphological characters.

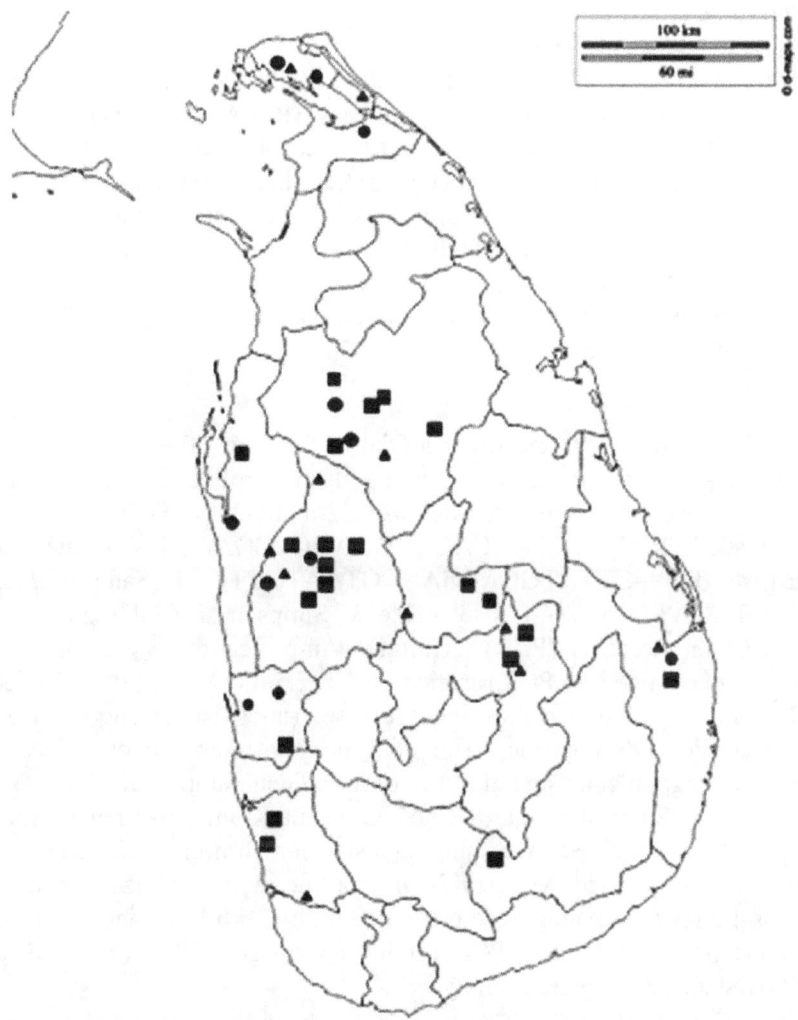

Fig. 1. The map depicting the locations where specimens were collected for the present study. ● *N. pubescens*-white (11 populations), ■ *N. pubescens*- pink (21 populations), ▲ Deep purplish red flowered *Nymphaea* (11 populations)

Morphological studies

Vegetative characters such as leaf shape, length, petiole diameter and reproductive characters such as flower size, petal and sepal length, number of stamens were studied either with the naked eye, under a dissecting microscope or under a stereo microscope (Leica, 10446322, 2X WD). Five individuals from each population were studied in detail where the measurement was averaged. The mean value for up to three measurements of each character was recorded for each specimen. Special attention was paid to characters with distinct variations. Colour of the lower and upper surfaces of the leaf and petal, stamens, stigmatic segments, and petiole were determined using the Royal Horticulture Society Colour Chart (RHS Colour Chart 2001). Nineteen quantitative and 21 qualitative characters were coded for 50 representatives of *N. pubescens* including both white and pink flowered populations and deep purplish red flowered *Nymphaea* species. All qualitative characters were expressed quantitatively by giving a coding value, to avoid misrepresenting the possible range of variation (Stevens, 1991).

Morphometric analyses

The multivariate statistical analysis was carried out using the PAST - Paleontological Statistics program version 2.17 (Hammer *et al.,* 2001). Cluster analysis, Principal Component Analysis (PCA) and Principal Coordination Analysis (PCoA) were performed. The cluster solution was selected from the best suitable similarity measure method and the algorithm; Gower similarity measure and 'Paired group' option (UPGMA), which produced the highest Co-phenetic correlation value of 0.933 over the other similarity distance methods and algorithms. Similarity Percentage Analysis (SIMPER) was performed to obtain overall average dissimilarity levels of the groups. Other than the SIMPER, PCA loadings were also used to rank characters regarding their contribution for the separation of clusters, and thereby to find the best vegetative characters to differentiate between the two species.

DNA studies

Total genomic DNA was extracted using a Qiagen DNAeasy Plant Mini kit, from fresh leaf materials from three selected samples representing deep purplish flowered *Nymphaea* , white flowered and pink flowered *N. pubescence.* matK (matK-390f 5'-CGATCTATTCATTCAA TATTTC-3', and matk-1326r 5' –TCTAGCACACGAAAGTCGAAGT- 3') (Cuenoud *et al.,* 2002) and *psbA-trnH* [*psbA*-F 5' –GTTATGCATGAACGTAATGCTC- 3' (Sang *et al.,* 1997), *trnH-R* 5'- CGCGCATGGTGGATTCACAATCC-3' (Tate & Simpson, 2003)] regions were amplified using Polymerase Chain Reaction (PCR) technique. Amplifications were carried out in 50 μL reaction solutions that contained 1× PCR reaction buffer, 2.5mM $MgCl_2$, 0.2 mM deoxynucleotide triphosphate (dNTPs), 0.2 μM each forward and reverse primer, 1 U of Taq DNA polymerase and 0.75–1.5 μL unquantified DNA extract. The PCR program was run on a Techne- Flexigene Thermal Cycler. The program consisted of 3 min of initial denaturation at 94°C, 35 cycles of 30 S denaturation at 94°C, 30 S annealing at 48°C/ 57°C for matK and *psbA-trnH* respectively, 1 min primer extension at 72°C, followed by a final extension for 10 min at 72°C. PCR products were run on a 1% agarose gel stained with ethidium bromide, and visualized on a UV table. The molecular mass of the resulted bands were estimated with a 1kb DNA ladder and confirmed the amplification of the primer. Obtained PCR products were submitted for sequencing reactions using Applied Biosystems, 3500 genetic analyzer.

Consensus for resulted sequences of forward and reverse primers was compiled using Bioedit version 7.1.11 and edited visually. Sequences deposited in the GenBank, for the *matK* gene region for the two taxa by other literatures were extracted using a BLAST (Basic Local Alignment Search Tool). ClustalW multiple sequence alignment was also used for Sequences alignment and comparison other than the BLAST.

Results and Discussion

The list of characters that were studied in detail together with their character states is given in the Table 1. The UPGMA dendrogram (co-phenetic correlation coefficient = 0.933) resolved two discrete clusters (denoted as cluster A and B), which separated respectively at approximately 0.35

Table 1. List of characters together with their character states.

Character	Character states
Diameter of the receptacle	mm
Receptacle height	cm
Flower colour (inner colour of petal)	White/Yellow/ Pink/ Deep purplish red
Flower colour (outer colour of petal)	White/ Yellow/ Pink/ Deep purplish red
Number of petals	
Petal length (outer petals)	cm
Petal width (maximum) (outer petals)	At the broadest point in cm
Petal shape	Linear-lanceolate/ Ob-lanceolate
Number of veins per petals	
Petal base	More or less widen into rectangular shape
Petal apex-shape and angle	Acute/ Obtuse
Number of stigmatic segments	
Number of sepals	Always 4 in number
Sepal length	cm
Sepal width (maximum)	cm
Sepal shape	Linear-lanceolate/ Ob-lanceolate
Sepal apex – shape and angle	Acute/ Obtuse
Number of stamens	
Stamen colour	Yellow/Red
Stigmatic segments colour	Yellow/ Crimson-red colour
Pedicel diameter	At the end of the receptacle end in cm
Pedicel shape	Round/Slightly flat/Oval
Pedicel shape in cross section	No. of lacunae
Petiole – cross section	No. of lacunae
Leaf size	Length/Width in cm
Leaf shape	Round/ Ellipsoid
Leaf length	Apex to base in cm
Leaf width	Across the mid rob in cm
Length/Width ratio	
Lamina colour (upper)	Dark green/Light green/Green
Lamina colour (lower)	Brownish-red/ Purple/Green
Leaf margin	Dentate/ Strongly dentate
Leaf venation (lower) number of veins	14 or less / Over 14
Leaf venation (lower) pattern	Prominent/ Not prominent
Leaf hairs (lower)	Long hairs/ Short hairs
Leaf apex	Division present/ Division absent
Petiole diameter	cm
Shape of the petiole	Round/Oval/ Irregular shape
Petiole colour	Yellowish-white/ Reddish-brown/ Green
Pedicel colour	Dark-green/ Brown/ Brownish red
Hairs on the petiole	Present/ Absent

distance units. The OTUs within each cluster grouped together closely, with none of them exceeding a distance of more than 0.7 units within any given cluster (Fig. 2). The scatter plot that resulted from the PCoA is given in Fig. 3 (Transformation component, C = 2). The first four (principal) eigenvalues recovered from the PCoA (1.3985, 0.2106, 0.1268, and 0.0878) accounted for 71.17% of the total variance (54.53%, 8.27%, 4.95%, and 3.43% respectively). A plot of the first and second coordinates (which provided the greatest separation of OTUs) returned a result similar to that obtained by the Cluster Analysis. Here the PCoA also resolved two discrete clusters, with each corresponding exactly to one of the clusters indicated by the UPGMA dendrogram.

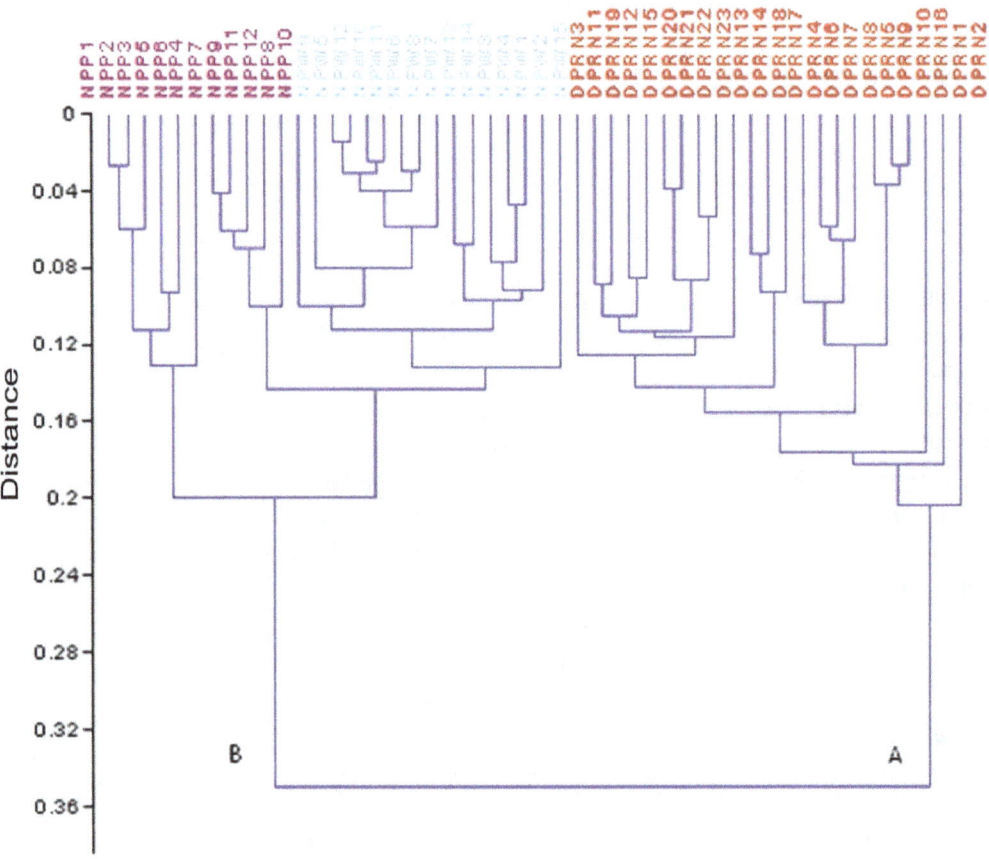

Fig. 2.   Dendrogram that resulted from morphometric analysis showing the clearly separated groups of *N. pubescens* and deep purplish red flowered *Nymphaea* species. Deep purplish red flowered *Nymphaea* species – DPRN, *N. pubescens* (white) – NPW and *N. pubescens* (pink) – NPP.

According to the results of PCA loading and SIMPER analysis, the number of stamens, leaf length, leaf width, number of petals and number of stigmatic segments are the highly contributed quantitative characters while lamina colour (both upper and lower), leaf venation pattern, petiole colour and stamen colour are the highly contributed qualitative characters for the separation. The character variation of the highly contributing six quantitative characters (number of stamens, leaf width, leaf length, number of stigmatic segments, number of petals and leaf size) between the two major groups; identified in analysis is given in Fig. 4 as box plots.

The morphometric analysis identifies two main phenetic groups A and B that corresponds to the deep purplish red flowered *Nymphaea* congregated in the cluster A while the cluster B corresponds to *N. pubescens*. Cluster B further branched at a distance of 0.2, where one group encompassed only pink flowered *N. pubescens* while the larger group consisted of both the white and pink flowered *N. pubescens*. Similarly the scatter plot obtained by PCoA clearly supports to the clustering of the populations into two major phenetic groups as A and B as recognized by the cluster analysis with non-overlapping distribution and the overlapped scattering of the members in group B. The detailed study of the characters of the members of the deep purplish red flowered *Nymphaea* and the character comparison with literature, Conard (1905) and La-ongsri *et al.* (2009), confirmed the identity of the group as *N. rubra* Roxb. ex Andrews, a species that has not been recorded before as occurring in the island.

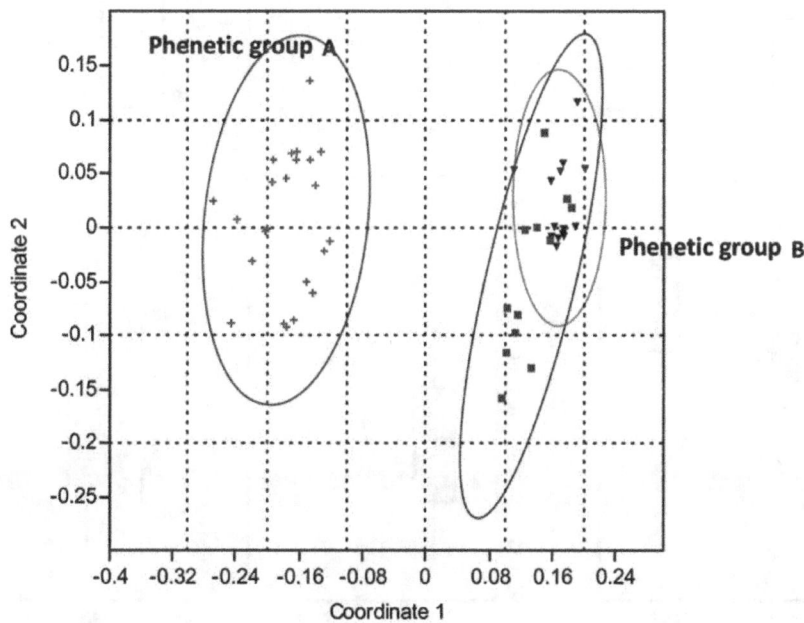

Fig. 3. Scatter plot at 95% ellipse level with eigenvalue scale obtained from PCoA.

DNA sequence analysis

The obtained sequence length of the *matK* and *psbA-trnH* gene regions were between 916-917 bp and 541-555 bp for deep purplish red flowered *Nymphaea* and *N. pubescens* respectively. The similarity percentage comparison of the obtained sequences of both *N. pubescens* (white and pink), with deep purplish red flowered *Nymphaea* with alignment scores obtained from BLAST search are given in Table 2. There were no sequence data deposited in the GenBank for both taxa for the *psbA-trnH* gene region.

Comparison of the obtained sequences with the GenBank (*N. rubra* - Acc. No. HQ592335.1) (Jeremy *et al.*, 2010) gave a 99% similarity for the deep purplish red flowered *Nymphaea* for the *matK* gene region with *N. rubra* while it gave a 96% similarity for *N. pubescens* (Acc. No. FJ597753.1) (Jeremy *et al.*, 2010). Comparison of the BLAST sequence divergence between *N. pubescens* and deep purplish red flowered *Nymphaea* for the sequences that were obtained in the study indicated only a 95% similarity for *matK* gene region existed between the two, while only 92% similarity for *psbA-trnH* gene region was indicated. Further, the comparison of both white and pink flowered *N. pubescens* sequences with the GenBank gave a 99% similarity match with *N. pubescens* (Acc. No. FJ597753.1).

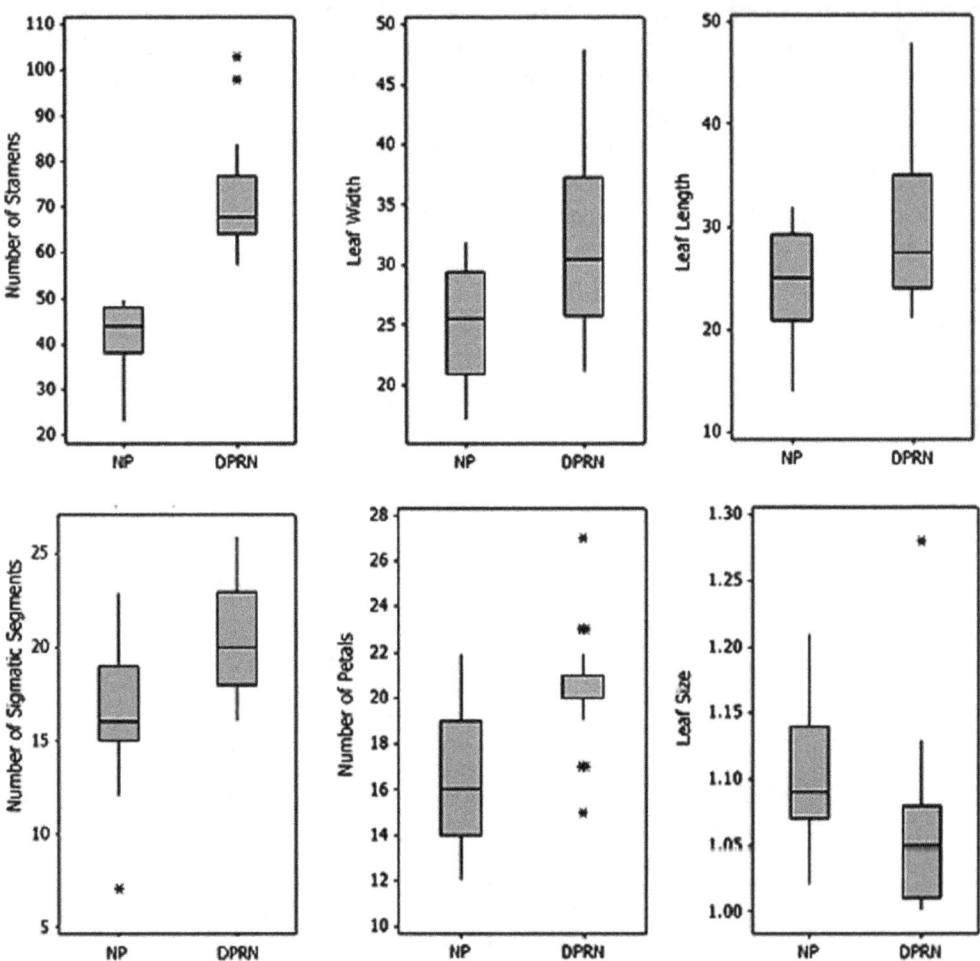

Fig. 4.   Box-plots of the six highly contributing quantitative characters between the two major groups identified in analysis [*N. pubescens* (NP), and Deep purplish red flowered *Nymphaea* (DPRN) species].

Table 2. Percentage similarity obtained from the comparisons of studied sequences and the GenBank accession, using BLAST.

Compared sequences	Blast score	
	matK	*trnH-psbA*
Deep purplish red *Nymphaea* (DPRN) *vs. N. pubescens* white (NPW)	95%	92%
Deep purplish red *Nymphaea* (DPRN) *vs. N. pubescens* pink (NPP)	98%	92%
N. pubescens white (NPW) vs. *N. pubescens* pink (NPP)	99%	99%
Deep purplish red *Nymphaea* (DPRN) *vs. N. rubra* (HQ592335.1)	99%	NA
N. pubescens white (NPW) *vs N. rubra* (HQ592335.1)	96%	NA
N. pubescens pink (NPP) *vs N. rubra* (HQ592335.1)	98%	NA

BLAST search results, while indicating that the deep purplish red flowered *Nymphaea* is a different taxa from the native *N. pubescens*, further confirms its identity as *N. rubra*. Comparison of the sequences using ClustalW, mismatches accounted for 6 point mutations and 17 insertion/

deletion (INDELS) events (single base pair) observed in the alignment for the two sequences of *matK* for the two species, *N. rubra* and white *N. pubescens* obtained in the present study while 4 point mutations and 3 INDELS (4, 5, and 17 bp length) were encountered for *psbA-trnH* sequence alignment. When compare between pink and white *N. pubescens,* there were only 2 gaps observed for *matK* and 5 gaps observed for *psbA-trnH* sequence. INDELS occurred in both *matK* and *psbA-trnH* gene regions, where the most informative was in the *psbA-trnH* region.

The results of both multivariate statistical analyses and the molecular sequences comparison have supported the recognition of the deep purplish red *Nymphaea* as a different species from *N. pubescens.* The detailed comparison of morphological characters has identified the species as *N. rubra* Roxb. ex Andrews while the molecular sequence comparison has further confirmed its identity.

According to the literature *N. rubra* could be easily separated from *N. pubescens* from the close examination of floral and leaf characters. Average dissimilarity value for the separation of *N. rubra* (DPRN) from white flowered *N. pubescens* (NPW) and pink flowered *N. pubescens* (NPP) is 15.31. According to the results, characters such as number of stamens, leaf length and width, and number of petals are good quantitative characters for delimitation of these two species while lamina colour (both upper and lower), leaf venation pattern, petiole and stamen colour are the highly contributed qualitative characters. Even though the cluster encompassing *N. pubescens* (Cluster B) initially separates a few individuals with pink flowers in a separate cluster, the remaining group includes both individuals with pink and white flowers once again separating the pink flowered into a sub-cluster. The overall dissimilarity value between the members of the two pink and white flowered groups within *N. pubescens* is 10.14 according to the SIMPER analysis while the gap was less than 0.2 in distance units (Fig. 2). Further in both BLAST and ClustalW sequence alignments for both pink and white flowered *N. pubescens,* the two sequences for both gene regions, showed a very high sequence identities with zero or very few mismatches, implying that they are just two color variations of the same species.

Table 3. Comparison of distinct morphological characters between *Nymphaea rubra* and *Nymphaea pubescens.*

Characters	*Nymphaea rubra*	*Nymphaea pubescens*
Flower size	Large (35 - 44 cm)	Small (28 - 36 cm)
Petal colour	Deep purplish red	White or light pink
	The colour intensity from apex to base on both adaxial and abaxial surfaces is uniform	The colour intensity is not uniform on both sides of the petal, showing a gradual fading from apex to base
Stamen - colour	Cinnabar red	Yellow
Stigmatic segments - colour	Crimson red	Yellow
Leaf size	Large (25 - 48 cm)	Small (25 - 30 cm)
Leaf shape	Orbicular	Ovate- orbicular
Leaf colour -adaxial	Bronzy red while young, turning dark green with age	Green
Leaf colour -abaxial	Dark purple colour	Brown
Venation	Over 9 pairs of very prominent secondary veins	7-9 prominent secondary veins

A comparison of the characters between the two species is given in Table 3 and Fig. 5, and an identification key for the Sri Lankan *Nymphaea* is given below.

1. Leaves pubescent beneath with many short hairs, margin sharply dentate-mucronate; stamens without a tongue-shaped appendage beyond the anther or appendage very short   2
-   Leaves glabrous, margin entire to dentate with blunt teeth; stamens with a tongue-shaped appendage beyond the anther to 5 mm long   *N. nouchali*
2. Leaf abaxial surface brown, venation pattern less prominent, petiole light green; flowers white, yellowish white or pink; stamens short, yellow; stigmatic surface yellow   *N. pubescens*
-   Leaf abaxial surface dark purple, venation pattern very prominent, petiole reddish; flowers deep purplish red; stamen long, cinnabar red; stigmatic surface crimson red   *N. rubra*

All Water-lily species occur together in large water bodies in both dry and wet zones of the country. However, in many instances *N. rubra* occurs towards the center of the deep waters in isolation.

Fig. 5. Deep purplish red flowered *Nymphaea* species identified as *N. rubra* during the present study (A). Flowers of *N. pubescens,* pink (B) and white (C). Leaf upper surfaces of *N. pubescens* (left) and *N. rubra* (right) (D) Cinnabar red colour stamens of *N. rubra* (E) and yellow colour stamens of *N. pubescens* (F).

The study has resulted in adding a new member to the genus *Nymphaea* in Sri Lanka enriching the islands biodiversity. *Nymphaea* is a taxonomically difficult group with many natural and man-made hybrids occurring in the nature and therefore, morphological features alone are not sufficient in confirming the identity. The present study provides additional reference sequences for both *N. pubescens* (Pink and white flowered groups) and *N. rubra* as well as a new gene region, *psbA-trnH* for reference.

Acknowledgements

Financial assistance provided by the National Science Foundation (NRB/2011/RG/03) is gratefully acknowledged. Authors wish to thank Menaka Ariyarathne, N. Shanjayan and all others who helped in field collections.

References

Biswal, Devendra K., Manish Debnath, Shakti Kumar, and Pramod Tandon. 2012. Phylogenetic reconstruction in the Order Nymphaeales: ITS2 secondary structure analysis and in silico testing of maturase k (matK) as a potential marker for DNA bar coding. BMC Bioinformatics **13** (17): S26.

Conard, H.S. 1905. The waterlilies: a monograph of the genus *Nymphaea*. Publications of Carnegie Institute of Washington, Washington, USA, 279 pp.

Cuenoud, P., Savolainen, V. and Chatrou, L.W. 2002. Molecular phylogenetics of Caryophyllales based on nuclear 18S rDNA and plastid rbcL, atpB, and matK DNA sequences. Am. J. Bot. **89**: 132-144.

Dassanayake, M.D. 1996. Nymphaeaceae. *In*: Dassanayake, M.D. & Clayton, W.D. (Eds.), A Revised Handbook to the Flora of Ceylon. Oxford & IBH Publ. Co. Pvt., Ltd., New Delhi, India, pp. 289-292.

de Vlas, J. and de Vlas-de Jong, J. 2008. Illustrated field guide to the flowers of Sri Lanka. Mark Booksellers and Distributors (Pvt) Ltd., Kandy, Sri Lanka. 179 p.

Gupta, P.P. 1980. Cytogenetics of aquatic ornamentals VI. Evolutionary trends and relationships in the genus *Nymphaea*. Cytologia **45**: 307-314.

Hammer Ø., Harper D.A.T. and Ryan P.D. 2001. PAST: Paleontological statistics software package for education and data analysis. Palaeontol. Electronica **4** (1): 1–9. http://palaeo-electronica.org/2001_1/past /issue1_01.htm. Retrieved on 25 August 2014.

Hossain, A., Kabir, G., Ud-deen, M. M., and Alam, A. M. S. 2007. Cytological studies of *Nymphaea* species available in Bangladesh. J.Bio-Science **15**: 7-13.

Jaime, B.B., Alejandro, N., Yolanda, H.O. and Judith, M.G. 2000. Comparative seed morphology of Mexican *Nymphaea* species. Aquatic Botany **68**: 189-204.

Jeremy, D., Suman, K., Satyawada, R.R. and Pramod, T. 2010. Molecular phylogenetics and the taxonomic reassessment of four Indian representative of the genus *Nymphaea*. Aquatic Botany **93**: 135-139.

La-ongsri, W., Trisonthi, C. and Balslev, H. 2009. A synopsis of Thai Nymphaeaceae. Nordic Journal of Botany **27**: 97-114.

Mitra, R.L. and Subramanyam, K. 1982. Is *Nymphaea rubra* Roxb. Ex. Andrews an apomict? Bull. Bot. Surv. India **24**: 83-86.

Polina, A.V. and Alexy, B.S. 2007. Morphological variation of *Nymphaea* (Nymphaeaceae) in European Russia. Nordic Journal of Botany **25**: 329-338.

Sang, T., Crawford, D.J. and Stuessy, T.F. 1997. Chloroplast DNA phylogeny, reticulate evolution and biogeography of Paeonia (Paeoniaceae). Am. J. Bot. **84**: 1120–1136.

Stevens, P. F. 1991. Character states, morphological variation, and phylogenetic analysis: a review. Systematic Botany **16**: 553-583.

Tate, J.A. and Simpson, B.B. 2003. Paraphyly of Tarasa (Malvaceae) and diverse origins of the polyploid species. Systematic Botany **28**: 723 –737.

12

APPLICATION OF INTERNAL TRANSCRIBED SPACER OF NUCLEAR RIBOSOMAL DNA FOR IDENTIFICATION OF *ECHINOPS MANDAVILLEI* KIT TAN

FAHAD M.A. AL-HEMAID, M. AJMAL ALI[1], JOONGKU LEE[2], GÁBOR GYULAI[3]
AND ARUN K. PANDEY[4]

Department of Botany and Microbiology, College of Science, King Saud University, Riyadh 11451, Saudi Arabia

Keywords: Echinops mandavillei; Asteraceae; ITS; nrDNA; Endemic; Saudi Arabia.

Abstract

The present study explored the use of internal transcribed spacers (ITS) sequences (ITS1-5.8S-ITS2) of nuclear ribosomal DNA (nrDNA) for identification of *Echinops mandavillei* Kit Tan, an endemic species to Saudi Arabia. The sequence similarity search using Basic Local Alignment Search Tool (BLAST) and phylogenetic analyses of the ITS sequence of *E. mandavillei* Kit Tan showed high level of sequence similarity (98%) with *E. glaberrimus* DC. (section *Ritropsis*). The novel primary sequence and the secondary structure of ITS2 of *E. mandavillei* could have a potential use for molecular genotyping.

Introduction

The genus *Echinops* L. belonging to the subtribe Echinopsinae of Cynareae, of the family Asteraceae comprise about 120 species (Vidović, 2011), and distributed in tropical Africa, the Mediterranean basin, temperate regions of Eurasia, Central Asia, Mongolia and North-eastern China, with the maximum number of species occurring in the Caucasus and the Middle East (Susanna and Garcia-Jacas, 2007). The genus received considerable interest for establishing natural groups with infrageneric classification (Sánchez-Jiménez *et al.*, 2010). Morphological characters, like the pappus, which is a key taxonomic character of Cynareae, the type and density of indumentum on stems, leaf shapes and phyllaries are considered least significance in dissemination of *Echinops* species (Mozaffarian, 2006; Sánchez-Jiménez *et al.*, 2010). In Saudi Arabia, there are nine *Echinops* species, *viz. E. abuzinadianus* Chaudhary, *E. erinaceus* Kit Tan, *E. glaberrimus* DC., *E. hystrichoides* Kit Tan, *E. macrochaetus* Fresen., *E. mandavillei* Kit Tan, *E. sheilae* Kit Tan, *E. viscosus* DC. and *E. yemenicus* Kit Tan. Of them, *E. abuzinadianus, E. mandavillei* and *E. sheilae* are endemic to Saudi Arabia, while remaining species have been reported from different geographic locations of Arabian Peninsula. *E. mandavillei* was reported to occur in Dahna, Summan and Nafud sands (Chaudhary, 2000).

The DNA sequence technology provides series of new data for molecular phylogeny and DNA barcoding which has now-a-days changed the paradigm of species identification (Ali and Choudhary, 2011; Ali *et al.*, 2014). From the first report of the utility of the internal transcribed spacers (ITS) sequence of nuclear ribosomal DNA (nrDNA) in plants (Baldwin, 1992), it has been

[1]Corresponding author. E-mail: majmalali@rediffmail.com
[2]International Biological Material Research Center, Korea Research Institute of Bioscience and Biotechnology, Daejeon 305 806, South Korea
[3]Institute of Genetics and Biotechnology, St. István University, Gödöllo H-2103, Hungary
[4]Department of Botany, University of Delhi, Delhi 110007, India

extensively used to distinguish even very closely related species (Chen *et al.*, 2010; Yao *et al.*, 2010). Moreover, in the last two decades, the ITS sequence technology has gained much attention, along with the smartest genes available for the molecular phylogeny and taxonomy (Ali *et al.*, 2013).

The ITS sequence technology has been used for molecular phylogeny of *Echinops* (Garnatje *et al.*, 2005), and series of other genera of Cynareae (Susanna *et al.*, 1999; Vilatersana *et al.*, 2000; Wang *et al.*, 2005, 2007; Hidalgo *et al.*, 2006); however, these studies did not include systematics of *Echinops* species occurring in Saudi Arabia. Hence, the present study aims to establish molecular signature of *Echinops mandavillei* Kit Tan based on ITS sequence of nrDNA.

Materials and Methods

Plant materials:

The leaf material of *Echinops mandavillei* Kit Tan was collected from herbarium specimen (Saudi Arabia, Al-Nafud, 29.4'N, 39.58'E, 5 May 1985, H.O. Al-Hassan 195) housed at National Herbarium and Genebank, National Agriculture and Animal Resources Research Centre, Riyadh, Saudi Arabia (RIY). The taxonomic identification of specimen was confirmed with the aid of Flora of Saudi Arabia (Chaudhary, 2000).

ITS sequences of 39 species of *Echinops* (Table 1) were retrieved from the GenBank database of NCBI (National Centre for Biotechnology Information; www.ncbi.nlm.nih.gov). *Brachylaena discolor* DC., from the tribe Tarchonantheae Kostel and *Cardopatium corymbosum* (L.) Pers. from the subtribe Cardopatiinae Less. were chosen as outgroups (Table 1) according to previous report based on molecular characters (Susanna *et al.*, 2006; Sánchez-Jiménez *et al.*, 2010).

Table 1. List of *Echinops* species used in the present study along with accession numbers.

Taxa	Accession number
Ingroup	
sect. ***Acantholepis*** (Less.) Jaub. & Spach	
1. *Echinops acantholepis* Jaub. & Spach	AY8262223
sect. ***Chamaechinops*** Bunge	
2. *E. fastigiatus* Kamelin & Tscherneva	GU116503
3. *E. humilis* M. Bieb	GU116514
4. *E. integrifolius* Kar. & Kir.	GU116517
sect. ***Echinops***	
5. *E. arachniolepis* Rech. f.	GU116486
6 *E. dahuricus* Fisch.	GU116493
7. *E. freitagii* Rech. f.	GU116504
8. *E. kotschyi* Boiss.	GU116520
9. *E. latifolius* Tausch	GU116521
10. *E. nizvanus* Rech. f.	GU116530
11. *E. parviflorus* Boiss. & Buhse	GU116533
12. *E. przewalskyi* Iljin	GU116535
13. *E. ritrodes* Bunge	GU116539
14. *E. setifer* Iljin	GU116540
15. *E. sphaerocephalus* L.	GU116541
16. *E. spiniger* Iljin	GU116542
17. *E. transcaucasicus* Iljin	GU116546

Table 1 contd.

Taxa	Accession number
sect. *Hamolepis* R. E. Fr.	
18. *E. hoehnelli* Schweinf	GU116506
sect. *Hololeuce* Rech. f.	
19. *E. hololeucus* Rech. f.	GU116513
sect. *Nanechinops* Bunge	
20. *E. gmelini* Turcz.	GU116510
sect. *Oligolepis* Bunge	
21. *E. cephalotes* DC.	GU116487
22. *E. cornigerus* DC.	GU116552
23. *E. echinatus* Roxb.	GU116497
24. *E. ghoranus* Rech. f.	GU116508
25. *E. griffithianus* Boiss.	GU116512
26. *E. ilicifolius* Bunge	GU116516
27. *E. leucographus* Bunge	GU116522
28. *E. lipskyi* Iljin	GU116523
sect. *Phaeochaete* Bunge	
29. *E. longifolius* A. Rich	GU116524
sect. *Psectra* Endl.	
30. *E. strigosus* L.	AY5386532
sect. *Ritropsis* Greuter & Rech. f.	
31. *E. chardinii* Boiss. & Buhse	GU116490
32. *E. dichrous* Boiss. & Hausskn.	GU116495
33. *E. endotrichus* Rech. f.	GU116500
34. *E. gaillardotii* Boiss.	GU116507
35. *E. glaberrimus* DC.	GU116509
36 *E. mandavillei* Kit Tan	KJ187107
37. *E. orientalis* Trautv.	GU116532
38. *E. spinosissimus* Turra	HE687348
39. *E. tenuisectus* Rech. f.	GU116551
sect. *Terma* Endl.	
40. *E. exaltatus* Schrad.	GU116501
Outgroup	
41. *Brachylaena discolor* DC.	AY8262363
42. *Cardopatium corymbosum* (L.) Pers.	AY8262383

DNA isolation and amplification:
    Genomic DNA was extracted from 10 mg silica gel-dried leaves using the protocol of DNeasy Plant Mini kit (QIAGEN, Valencia, CA, USA). The ITS regions were amplified using the primers ITS1 and ITS4 as described by White *et al.* (1990). Double-stranded polymerase chain reaction (PCR) products were produced through 35 cycles of 95°C for 1 min, 48°C for 1 min and 72°C for 1 min, with a 10 min final extension cycle at 72°C. PCR products were purified with SolGent PCR Purification kit-Ultra (SolGent, Daejeon, South Korea), and forwarded to sequencing using the same primers, 2L BigDye, 1µl primer (20 pM), template DNA and purified water to reach a 10µl reaction volume. Cycle sequencing used was 25 cycles of 96°C for 10 s, 50°C for 5 s, and 60°C for 4 min.

DNA sequencing and data analysis:

DNA sequencing was performed by ABI Prism 377 automated DNA sequencer (Applied Biosystems, Foster City, CA, USA). Each sample was sequenced in the sense and anti-sense direction. The nucleotide sequences of both DNA strands were obtained and analyzed by Sequence Navigator (Perkin-Elmer/Applied Biosystems) to ensure accuracy of the base pair sequences. The sequence was submitted to GenBank (accession number KJ187107).

Sequence alignments were performed using CLUSTAL X, version 1.81 (Thompson *et al.*, 1997), and sequence alignments were subsequently adjusted manually using BioEdit (Hall, 1999). Gaps were treated as missing data in phylogenetic analyses. The maximum parsimony and neighbour-joining analyses with 1,000 bootstrap replicates (Felsenstein, 1985) were performed using PAUP* 4.0b10 (Swofford, 2002). The boundaries between ITS1, 5.8S and ITS2 gene were determined according to the ITS sequences of *Echinops* available in GenBank. The ITS2 database (http://its2.bioapps.biozentrum.uni-wuerzburg.de/) was used to predict the secondary structures (Koetschan *et al.*, 2012).

Results and Discussion

The ITS region (ITS1-5.8S-ITS2) of *Echinops mandavillei* Kit Tan sequenced in the present study was found 634 bp, where ITS1 region 252 bp (GC content 54%), 5.8S gene 164 bp (GC content 53%), and ITS2 region 218 bp (GC content 50%). The BLAST search of ITS sequence of *E. mandavillei* Kit Tan showed maximum identity (98%) with *E. glaberrimus* DC. Parsimony analysis of the entire ITS region resulted in 431 maximally parsimonious trees with consistency index of 0.691, homoplasy index of 0.459, and retention index of 0.763. The phylogenetic tree constructed by the present analyses shows *Echinops* to be monophyletic (bootstrap support 100%; Fig. 1). The tree also provides a clear resolution at the sectional level and the result confirms an earlier report (Sánchez-Jiménez *et al.*, 2010), and *E. mandavillei* Kit Tan nested within the clade of the section *Ritropsis* (Fig. 1). Figure 2 illustrates specific nucleotide differences between *E. mandavillei* Kit Tan and *E. glaberrimus* DC., in total seven SNPs (four nucleotides in ITS1 region, i.e. at the alignment position 11, 81, 226 and 234, and three nucleotides in ITS2 region, i.e. at the alignment position 4, 58 and 165) were observed.

Table 2. Loci of SNPs (single nucleotide polymorphism) ITS sequences of *E. mandavillei* compared to *E. glaberrimus*.

Region	Position in sequence alignment	*E. mandavillei* → *E. glaberrimus*
ITS1	11th	T → C
	81th	G → R
	226th	T → C
	234th	C → T
ITS2	4th	A → C
	58th	A → G
	165th	T → C

The secondary structures of ITS2 region of *E. mandavillei* Kit Tan and *E. glaberrimus* DC. were constructed and compared (Fig. 3 A-B), which contained a central ring (primary ring) and four helices. However, the two structures differed in the four helical regions, in stem loop numbers, sizes, position, and screw angle. On the basis of the ITS2 secondary structure, *E. mandavillei* Kit Tan could be discriminative from other species of the genus.

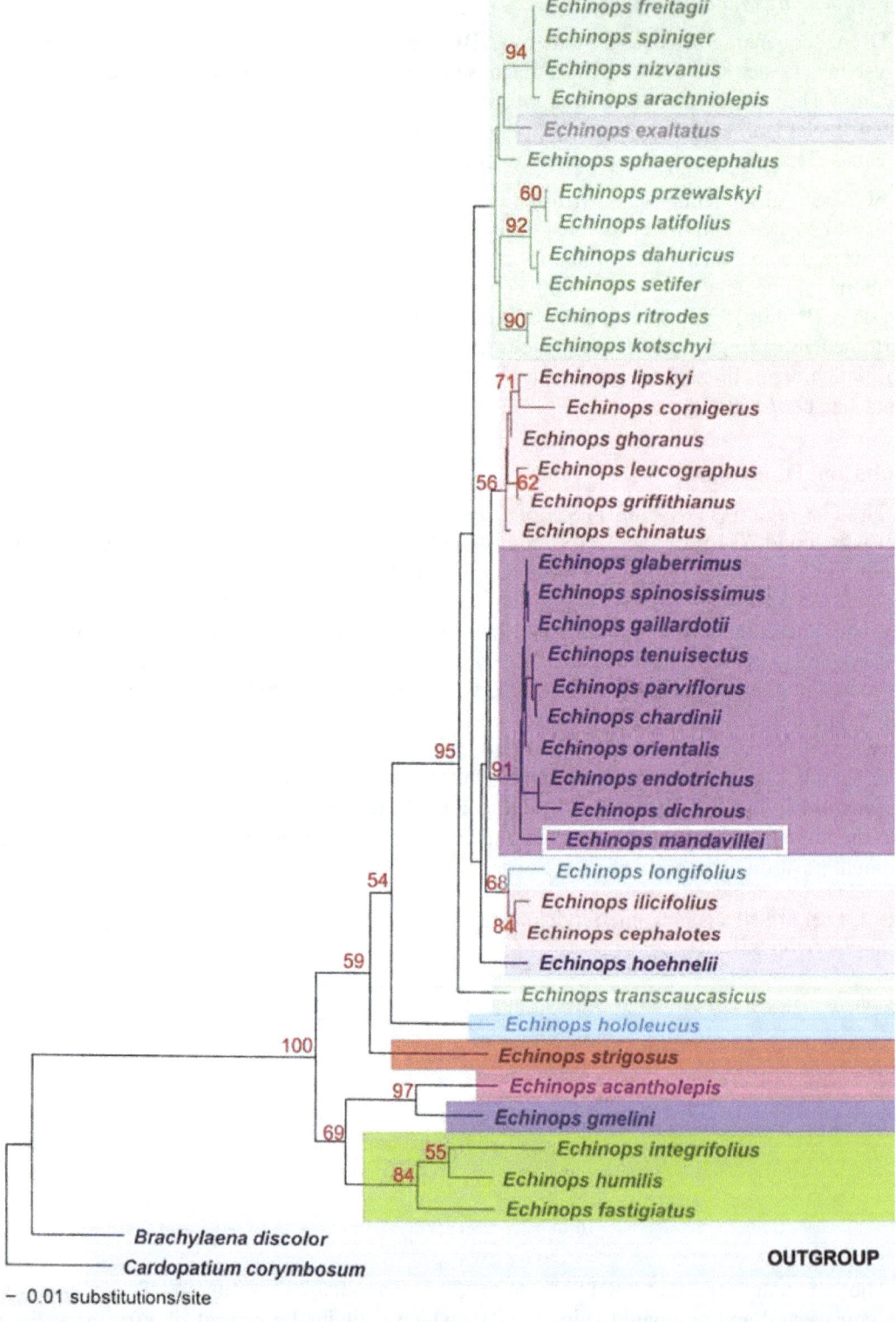

Fig. 1. Neighbour joining tree of *Echinops* species including *E. mandavillei* inferred from ITS sequences of nrDNA. Bootstrap values greater than 50% in 1,000 bootstrap replicates are shown above lines.

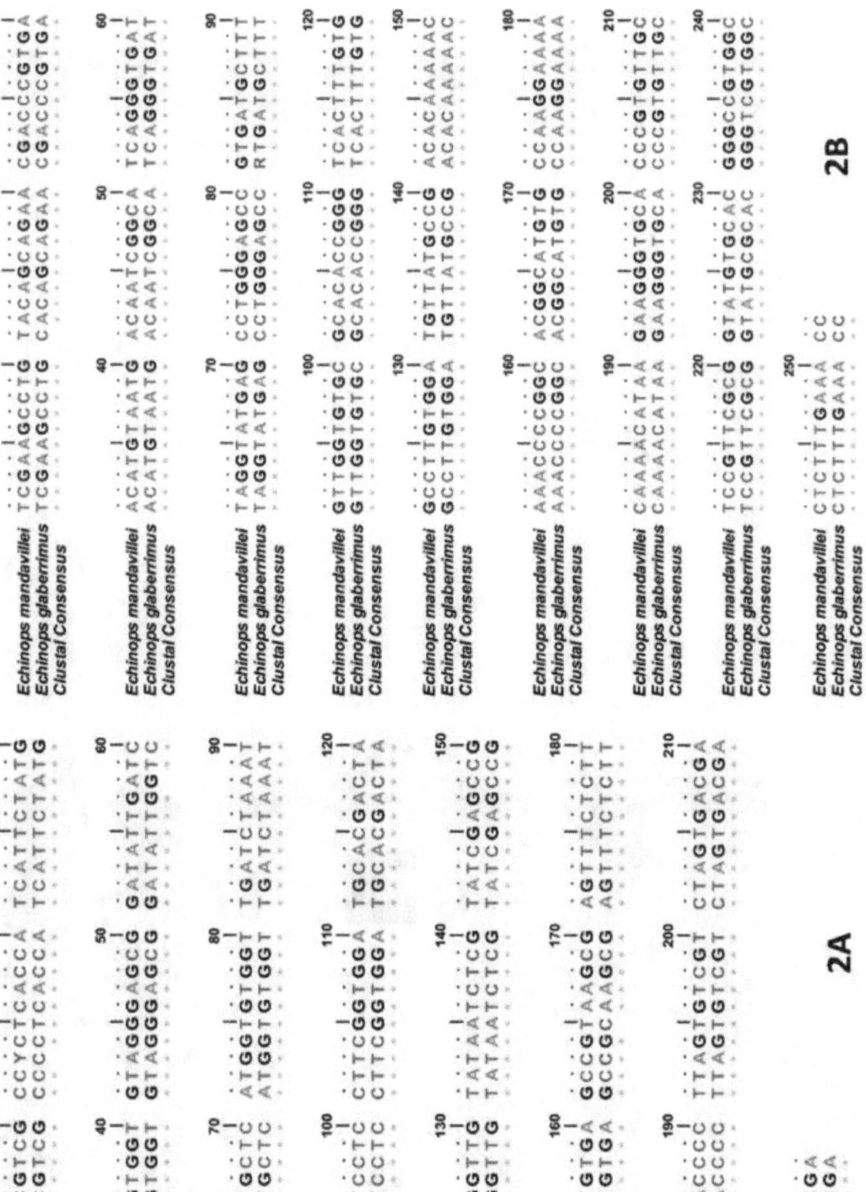

Fig. 2A. Alignments of ITS1 sequences of *E. mandavillei* compared to *E. glaberrimus*, B. Alignments of ITS2 sequences of *E. mandavillei* compared to *E. glaberrimus*. Gaps in clustal line indicate nucleotide differences.

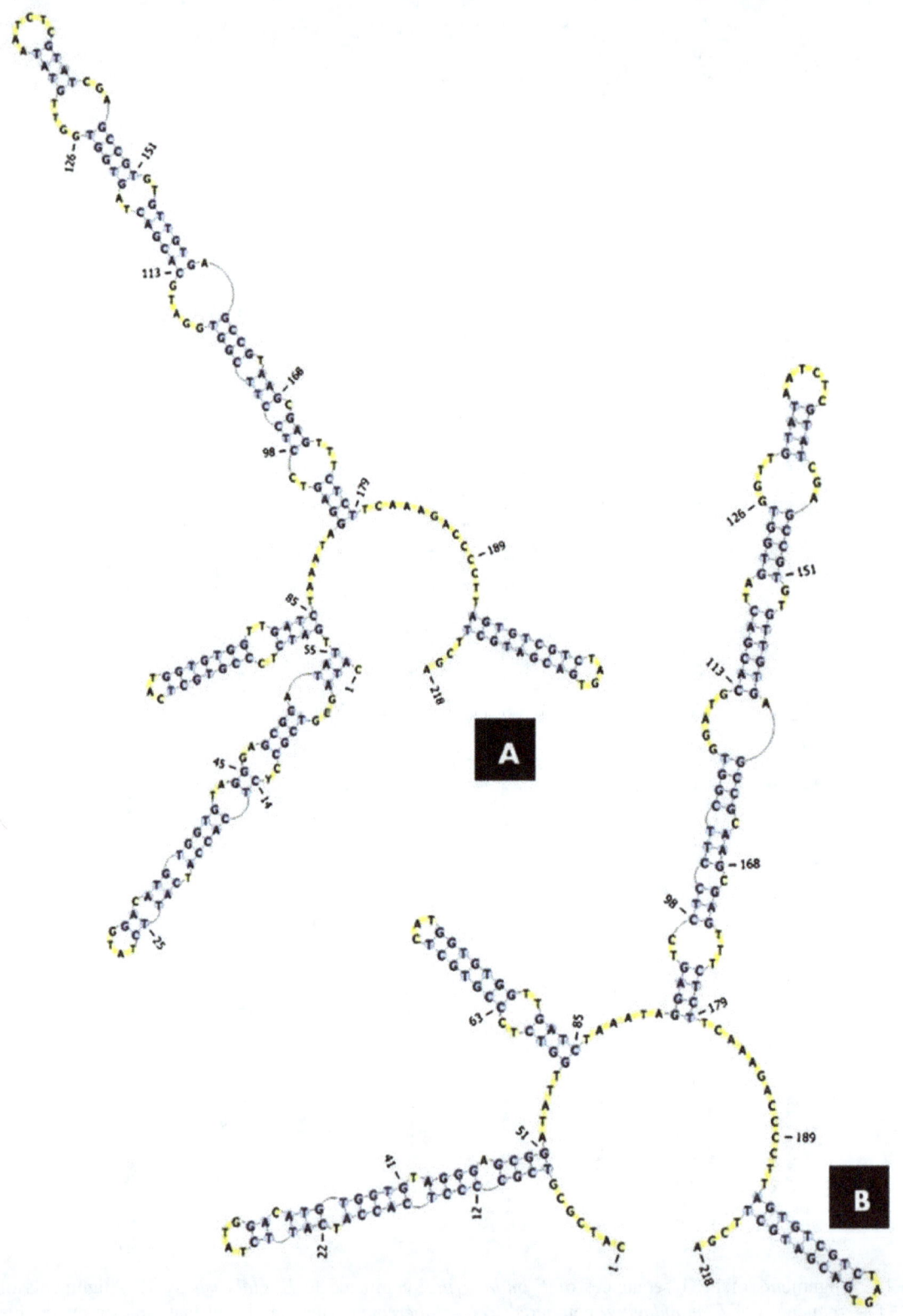

Fig. 3. The secondary structures of the ITS2 regions of *E. mandavillei* (A) compared to *E. glaberrimus* (B).

The morphological identification depends on sufficient experience and can easily be affected by the geographical environment and biocoenosis (Marcon *et al.*, 2005; Rai *et al.*, 2012). In contrast, DNA sequence is hardly influenced by environmental characteristics and developmental stages (Liu *et al.*, 2011); and therefore, the DNA barcoding may be an effective supplement to traditional/classical morphological methods (see Hebert *et al.*, 2003). The species identification using DNA barcodes has been successfully used across the algae, fungi, plants, and animals, hence; the DNA barcoding has now been proven useful in biodiversity assessment, biomonitoring, forensics, illegal trade of endangered species and their products, ecology, medicinal and poisonous plants and conservation genetics (see Hebert *et al.*, 2003; Fišer Pečnikar and Buzan, 2014; Ali *et al.*, 2014).

DNA barcoding efforts worldwide have resulted in the formation of the Consortium for the Barcode of Life (CBOL), and the Barcode of Life Database (BOLD), which contain more than 2.7 million records, with 2 million barcodes belonging to over 170,000 species (Ratnasingham and Hebert, 2007; BOLD Systems, 2013). The China Plant BOL Group has proposed that ITS1/ITS2 should be incorporated into the core of barcode for seed plants (Li *et al.*, 2011). In the present study, we supplied the ITS barcode of *E. mandavillei* Kit Tan which is new for GenBank databases. An increasing number of studies also suggest that DNA secondary structures are crucial for genomic stability and cellular processes, such as transcription (Bochman *et al.*, 2012; Salvi and Mariottini, 2012), and our study has also provided new data of *E. mandavillei* Kit Tan for this purposes.

Acknowledgements

The authors would like to extend their sincere appreciation to the Deanship of Scientific Research at King Saud University for the funding of this research through the Research Group Project No. RGP-VPP-195.

References

Ali M.A, Al-Hemaid, F.M., Choudhary, R.K., Lee, J., Kim, S.Y. and Rub, M.A. 2013. Status of *Reseda pentagyna* Abdallah & A.G. Miller (Resedaceae) inferred from combined nuclear ribosomal and chloroplast sequence data. Bangladesh J. Plant Taxon. **20**(2): 233-238.

Ali, M.A., Gábor, G., Norbert, H., Balázs, K., Al-Hemaid, F.M.A., Pandey, A.K. and Lee, J. 2014. The changing epitome of species identification- DNA barcoding. Saudi J. Biol. Sci. http://dx.doi.org/ 10.1016/j.sjbs.2014.03.003

Ali, M.A. and Choudhary R.K. 2011. India needs more plant taxonomists. Nature **471**: 37.

Baldwin, B.G. 1992. Phylogenetic utility of the internal transcribed spacers of nuclear ribosomal DNA in plants: an example from the Compositae. Mol. Phylogenet. Evol. **1**: 3-16.

Bochman, M.L., Paeschke, K. and Zakian, V.A. 2012. DNA secondary structures: stability and function of G-quadruplex structures. Nat. Rev. Genet. **13**: 770-780.

BOLD Systems 2013. BOLD Systems v3. http://www.boldsystems.org/ (Retrieved on 6 May 2014)

Chaudhary, S. 2000. *Echinops. In:* Chaudhary, S. (Ed.), Flora of the Kingdom of Saudi Arabia. Ministry of Agriculture and Water, National Herbarium, National Agriculture and Water Research Center, Riyadh, Saudi Arabia, Vol. **II** (3), pp. 194-199.

Chen, S., Yao, H., Han, J., Liu, C., Song, J., Shi, L., Zhu, Y., Ma, X., Gao, T., Pang, X., Luo, K., Li, Y., Li, X., Jia, X., Lin, Y. and Leon, C. 2010. Validation of the ITS2 region as a novel DNA barcode for identifying medicinal plant species. PLoS ONE **5**(1): e8613.

Felsenstein, J. 1985. Confidence limits on phylogenies: an approach using the bootstrap. Evolution **39**: 783-791.

Fišer Pečnikar, Z. and Buzan, E.V. 2014. 20 years since the introduction of DNA barcoding: from theory to application. J. Appl. Genet. **55**: 43-52.

Garnatje, T., Susanna, A., Garcia-Jacas, N., Vilatersana, R. and Vallès, J. 2005. A first approach to the molecular phylogeny of the genus *Echinops* L. (Asteraceae): sectional delimitation and relationships with the genus *Acantholepis* Less. Folia Geobot. **40**: 407-419.

Hall, T.A. 1999. BioEdit: A user-friendly biological sequence alignment editor and analysis program for Windows 95/98/NT. Nuc. Acids Symp. Ser. **41**: 95-98.

Hebert, P.D.N., Cywinska, A., Ball, S.L. and deWaard, J.R. 2003. Biological identifications through DNA barcodes. Proc. Biol. Sci. **270**: 313-321.

Hidalgo, O., Garcia-Jacas, N., Garnatje, T. and Susanna, A. 2006. Phylogeny of *Rhaponticum* (Asteraceae, Cardueae-Centaureinae) and related genera inferred from nuclear and chloroplast DNA sequence data: Taxonomic and biogeographic implications. Ann. Bot. **97**: 705-714.

Koetschan, C., Hackl, T., Müller, T., Wolf, M., Förster, F. and Schultz, J. 2012. ITS2 database IV: Interactive taxon sampling for Internal Transcribed Spacer 2 based phylogenies. Mol. Phylogenet. Evol. **63**: 585-588.

Li, D.Z., Gao, L.M., Li, H.T., Wang, H., Ge, X.J., Liu, J.Q., Chen, Z.D., Zhou, S.L., Chen, S.L., Yang, J.B., Fu, C.X., Zeng, C.X., Yan, H.F., Zhu, Y.J., Sun, Y.S., Chen, S.Y., Zhao, L., Wang, K., Yang, T. and Duan, G.W. 2011. Comparative analysis of a large dataset indicates that internal transcribed spacer (ITS) should be incorporated into the core barcode for seed plants. Proc. Natl. Acad. Sci. USA **108**: 19641-19646.

Liu, C., Liang, D., Gao, T., Pang, X., Song, J., Yao, H., Han, J., Liu, Z., Guan, X., Jiang, K., Li, H. and Chen, S. 2011. PTIGS-IdIt, a system for species identification by DNA sequences of the *psb*A-*trn*H intergenic spacer region. BMC Bioinformatics **12**: S4.

Marcon, A.B., Barros, I.C. and Guerra, M. 2005. Variation in chromosome numbers, CMA bands and 45S rDNA sites in species of *Selaginella* (Pteridophyta). Ann. Bot. **95**: 271-276.

Mozaffarian, V. 2006. A taxonomic survey of *Echinops* L. tribe Echinopeae (Asteraceae) in Iran: 14 new species and diagnostic keys. Iranian J. Bot. **11**: 197-239.

Rai, P.S., Bellampalli, R., Dobriyal, R.M., Agarwal, A., Satyamoorthy, K. and Narayana, D.A. 2012. DNA barcoding of authentic and substitute samples of herb of the family Asparagaceae and Asclepiadaceae based on the ITS2 region. J. Ayurveda Integr. Med. **3**: 136-140.

Ratnasingham, S. and Hebert, P.D.N. 2007. Bold: The Barcode of Life Data System (http://www.barcodinglife.org). Mol. Ecol. Notes **7**: 355-364.

Salvi, D. and Mariottini, P. 2012. Molecular phylogenetics in 2D: ITS2 rRNA evolution and sequence-structure barcode from Veneridae to Bivalvia. Mol. Phylogenet. Evol. **65**: 792-798.

Sánchez-Jiménez, I., Lazkov, G.A., Hidalgo, O. and Garnatje, T. 2010 Molecular systematics of *Echinops* L. (Asteraceae, Cynareae): A phylogeny based on ITS and *trn*L-*trn*F sequences with emphasis on sectional delimitation. Taxon **59**(3): 698-708.

Susanna, A. and Garcia-Jacas, N. 2007. The tribe Cardueae. *In*: Kadereit, J.W. and Jeffrey, C. (Eds), The families and genera of vascular plants, Vol. **8**, Flowering plants; Eudicots; Asterales. Heidelberg, Springer, pp. 135-158.

Susanna, A., Garcia-Jacas, N., Hidalgo, O., Vilatersana, R. and Garnatje, T. 2006. The Cardueae (Compositae) revisited: Insights from ITS, *trn*L-*trn*F, and *mat*K nuclear and chloroplast DNA analysis. Ann. Missouri Bot. Gard. **93**: 150-171.

Susanna, A., Garnatje, T. and Garcia-Jacas, N. 1999. Molecular phylogeny of *Cheirolophus* (Asteraceae: Cardueae-Centaureinae) based on ITS sequences of nuclear ribosomal DNA. Plant Syst. Evol. **214**: 147-160.

Swofford, D.L. 2002. PAUP* (v. 4.0b10). Phylogenetic analysis using parsimony (* and other methods). Sinauer Associates, Sunderland.

Thompson, J.D., Gibson, T.J., Plewniak, F., Jeanmougin, F. and Higgins, D.G. 1997. The CLUSTAL_X windows interface: Flexible strategies for multiple sequence alignment aided by quality analysis tools. Nucleic Acids Res. **24**: 4876-4882.

Vidović, B. 2011. A new *Aceria* species (Acari: Eriophyoidea) on *Echinops ritro* L. subsp. *ruthenicus* (M. Bieb.) Nyman (Asteraceae) from Serbia and a supplement to the original description of *Aceria brevicincta* (Nalepa 1898). Zootaxa **2796**: 56-66.

Vilatersana, R., Susanna, A., Garcia-Jacas, N. and Garnatje, T. 2000. Generic delimitation and phylogeny of the *Carduncellus-Carthamus* complex (Asteraceae) based on ITS sequences. Plant Syst. Evol. **221**: 89-105.

Wang, Y.J., Liu, J.Q. and Miehe, G. 2007. Phylogenetic origins of the Himalayan endemic *Dolomiaea*, *Diplazoptilon* and *Xanthopappus* (Asteraceae: Cardueae) based on three DNA regions. Ann. Bot. **99**: 311-322.

Wang, Y.J., Pan, J.T., Liu, S.W. and Liu, J.Q. 2005. A new species of *Saussurea* (Asteraceae) from Tibet and its systematic position based on ITS sequence analysis. Bot. J. Linn. Soc. **147**: 349-356.

White, T.J., Bruns, T., Lee, S. and Taylor, J. 1990. Amplification and direct sequencing of fungal ribosomal RNA genes for phylogenetics. *In:* Innis, M.A., Gelfand, D.H., Sninksky, J.J. and White, T.J. (Eds), PCR protocols: a guide to method and amplifications. Academic Press, San Diego, California, pp. 315-322.

Yao, H., Song, J., Liu, C., Luo, K., Han, J., Li, Y., Pang, X., Xu, H., Zhu, Y., Xiao, P. and Chen, S. 2010. Use of ITS2 region as the universal DNA barcode for plants and animals. PLoS ONE **5**(10): e13102.

13

MOLECULAR PHYLOGENY OF SAUDI ARABIAN *TETRAENA* MAXIM. AND *ZYGOPHYLLUM* L. (ZYGOPHYLLACEAE) BASED ON PLASTID DNA SEQUENCES

Dhafer Ahmed Alzahrani[1] and Enas Jameel Albokhari[2]

Department of Biological Sciences, Faculty of Science, King Abdulaziz University, Jeddah, Saudi Arabia.

Keywords: cpDNA; *rbc*L; *trn*L-F; Phylogeny; Saudi Arabia; *Tetraena*; *Zygophyllum*; Zygophyllaceae.

Abstract

In order to provide a basis for better understanding of phylogenetic relationships of Saudi Arabian *Tetraena* Maxim. and *Zygophyllum* L., 44 specimens representing seven taxa, were reconstructed based on chloroplast DNA data of *rbc*L and *trn*L-F. The combined chloroplast (*rbc*L and *trn*L-F) contributed more phylogenetically informative characters than in individual regions. Phylogenetic analysis of the combined chloroplast (*rbc*L and *trn*L-F) and in individual regions based on both of Maximum Parsimony and Bayesian criteria showed that the Saudi Arabian species of *Tetraena* and *Zygophyllum* were monophyletic. *Zygophyllum fabag* L. was nested in one clade with *Z. xanthoxylum* (Bunge) Engl. (Asian species), and all taxa of *Tetraena* were distributed in other clades.

Introduction

The widespread family Zygophyllaceae includes five subfamilies *viz.* Zygophylloideae, Tribuloideae, Seetzenioideae, Larreoideae and Morkillioideae (Sheahan and Chase, 2000; Beier *et al.*, 2003; Bellstedt *et al.*, 2008). The *Zygophyllum* L. and *Tetraena* Maxim. belong to Zygophylloideae along with *Fagonia* L., *Augea* Thunb., *Roepera* (A. Juss.) Engl. and *Melocarpum* (Engl.) Beier & Thulin (Beier *et al.*, 2003; Bellstedt *et al.*, 2008).

The only detailed examination of the systematics of *Zygophyllum* and *Tetraena* taxa have focused on morphological and anatomical characters (El-Hadidi, 1977, 1980; Boulos, 1978; Engler, 1931; Hosny, 1988; Hussein *et al.*, 2009; Ma and Zhang, 1990; Takhtajan, 1987; Thulin, 1993; Van Huyssteen, 1937; Van Zyl, 2000). In contrast, a few studies have used molecular markers to the phylogenetic relationships of the intergeneric of Zygophyllaceae (Sheahan and Chase, 1996, 2000; Beier *et al.*, 2003) or to infer the relationships within the genus *Zygophyllum* (Bellstedt *et al.*, 2008; Hammad and Qari, 2010). Sheahan and Chase (1996) studied the phylogenetic relationships of Zygophyllaceae based on morphology, anatomy and the *rbc*L DNA sequence. Sheahan and Chase (2000) investigated the phylogenetic relationships of 36 taxa of Zygophyllaceae including 15 species of *Zygophyllum* L. from Africa, Australia, and south western Asia using nucleotide sequences of the plastid gene *rbc*L and non-coding *trn*L-F and found *Zygophyllum* as polyphyletic. They showed that the *Zygophyllum fabago* L. (the type species of *Zygophyllum*) nested with another Asian species *Z. xanthoxylum* (Bunge) Engl., whereas *Z. simplex* L. placed in a strong clade with the genus *Tetraena* and other *Zygophyllum* species, *viz. Z. album* L. f., *Z. coccineum* L., *Z. cylindrifolium* Schinz and *Z. decumbens* Delile (the last three are

---

[1]Corresponding author. Email: dalzahrani@kau.edu.sa
[2]Department of Biological Sciences, Faculty of Applied Sciences, Umm Al-Qura University, Makkah, Saudi Arabia.

distributed in Saudi Arabia). The study indicated that *Tetraena* is nested within the large and variable *Zygophyllum* and reported that the *Z. simplex* is sister to *Tetraena*.

Beier *et al.* (2003) investigated the phylogenetic relationships of Zygophyllaceae using *trn*L plastid DNA sequences and morphological data for 43 species of Zygophylloideae including the genera *Zygophyllum*, *Fagonia*, *Augea* and *Tetraena* which represent most of the morphological and geographical variations in the subfamily Zygophylloideae. They reported that the subfamily Zygophylloideae is monophyletic, whereas the genus *Zygophyllum* is paraphyletic, since this genus was spontaneously distributed with the genera of *Augea*, *Tetraena* and *Fagonia*. Based on the results of this study, Beier *et al.* (2003) produced a new classification for genera *Tetraena* and *Zygophyllum*, and transferred 35 species from genus *Zygophyllum* to genus *Tetraena* as new combinations. Later, Bellstedt *et al.* (2008) assessed the phylogenetic relationships of 53 species of *Zygophyllum* in southern Africa employing the sequences of *rbc*L and *trn*L-F regions. They included the published sequences of the same genes for other species from different regions and the results supported the subdivision of the genus *Zygophyllum* into subgenera *Agrophyllum* and *Zygophyllum*. They found relatively similar results by conducting the same methods to study the relationships of *Zygophyllum* and *Tetraena* species (cpDNA sequences) and similar morphological characteristics (i.e. capsule dehiscence, seed attachment and the presence of spiral threads in the seed mucilage). These species are known from Africa and Asia. Bellestedt *et al.* (2008) did not agree with Beier *et al.* (2003) for the new classification of *Tetraena* and *Zygophyllum*. However, many authors agreed with this transfer and used the combinations proposed by Beier *et al.* (2003) as valid in their works, including Alzahrani (2017), Alzahrani and Albokhari (2017a, b), Azevedo (2014), Ghazanfar and Osborne (2015), Louhaichi *et al.* (2011), Mosti *et al.* (2012), Norton *et al.* (2009), Sakkir *et al.* (2012).

Tetraena is represented in Saudi Arabia by six species, two subspecies and six varieties, while genus *Zygophyllum* is represented by a single species, namely *Z. fabago* (Beier *et al.*, 2003; Alzahrani, 2017; Alzahrani and Albokhari, 2017a, b;). Saudi Arabian *Tetraena* and *Zygophyllum* have never been included in the published phylogenetic studies. The only two studies have used RAPD markers data to study genetic variation among and within populations of some Saudi Arabian *Zygophyllum* taxa (Al-Arjany, 2011; Hammad and Qari, 2010). Hammad and Qari (2010) studied the genetic diversity of 12 populations of *Zygophyllum coccineum*, *Z. album* and *Z. aegyptium* A.I. Hosny which were collected from various locations in Egypt and Saudi Arabia using RAPD markers employing five random primers. They found that *Zygophyllum coccineum* revealed higher levels of genetic variation and more unique alleles than the other species and *Z. aegyptium* is genetically closely related to *Z. album*   Later, Al-Arjany (2011) studied the molecular taxonomy of *Zygophyllum simplex* and *Z. migahidii* using of random PCR (RAPD) technology to analyse phylogenetic relationships between both species and found that these species are closely allied to each other. In the present study, phylogenetic relationships of 43 individual specimens of Saudi Arabian *Tetraena* and *Zygophyllum* species were reconstructed using combined DNA sequences data from the *rbc*L and the *trn*L-F regions.

Materials and Methods

Selection of ingroup and outgroup

Leaf material for 37 individual specimens of Saudi Arabian *Tetraena* representing six taxa were sampled in the field and from herbarium specimens listed in Tables 1 and 2. Collected specimens were deposited in KAUH (King Abdulaziz University Herbarium, Jeddah, Saudi Arabia). Sequenced data of the 10 *Tetraena* and *Zygophyllum* sequenced by Bellstedt *et al.* (2008) for the two regions (*rbc*L and *trn*L-F) were obtained from GenBank (Table 3). Three sequences of

Table 1. Accessions of Saudi Arabian *Tetraena* collected and used for the molecular study.

Sl. No.	Vr. No.	Taxa	Location	Coordinates	GenBank accession *rbcL*	GenBank accession *trnL*-F
1.	109	*Tetraena propinqua* ssp. *propinqua*	Shuaibah	20° 52' 23" N 39° 22' 16" E	MG664288	MG664319
2.	110	*T. alba* var. *alba*	Shuaibah	20° 52' 23" N 39° 22' 16" E	MG664308	MG664339
3.	111	*T. coccinea*	Shuaibah	20° 52' 23" N 39° 22' 16" E	MG664302	MG664333
4.	117	*T. coccinea*	North of Jeddah	21° 50' 23" N 39° 07' 05" E	MG664303	MG664334
5.	120	*T. coccinea*	South of Alleith	19° 56' 15" N 40° 31' 17" E	MG664304	MG664335
6.	128	*T. coccinea*	Between Rabigh and Yanbu	23° 59' 20" N 38° 16' 02" E	MG664305	MG664336
7.	130	*T. coccinea*	Umluj	24° 33' 20" N 37° 25' 23" E	MG664306	MG664337
8.	133	*T. propinqua*ssp.*propinqua*	Umluj	24° 59' 05" N 37° 17' 09" E	MG664289	MG664320
9.	137	*T. propinqua*ssp. *propinqua*	Umluj	24° 58' 19" N 37° 17' 03" E	MG664290	MG664321
10.	138	*T. alba* var. *arabica*	Umluj	24° 58' 19" N 37° 17' 03" E	MG664310	MG664341
11.	139	*T. alba* var. *alba*	Umluj	24° 58' 19" N 37° 17' 03" E	MG664309	MG664340
12.	142	*T. decumbens*	30 km South of Umluj	24° 45' 06" N 37° 19' 56" E	MG664307	MG664338
13.	143	*T. propinqua*sspmigahidii	WadiTarabah	23° 20' 10" N 12° 41' 47" E	MG664286	MG664317
14.	145	*T. propinqua*sspmigahidii	Al-Qaeid road - Hail	27° 41' 18" N 41° 44' 38" E	MG664287	MG664318
15.	146	*T. simplex*	Alnuqrah - PrinceAbdul AzizbinMuqrin road-Hail	27° 27' 27" N 41° 38' 59" E	MG664281	MG664312
16.	D1	*T. simplex*	Dhalam - Taif-Riyadh road	22° 12' 10" N 41° 24' 19" E	MG664280	MG664311
17.	D5	*T. propinqua*sspmigahidii	Alkhasrah- Taif-Riyadh road	23° 24' 59" N 43° 43' 27" E	MG664282	MG664313
18.	D7	*T. propinqua*sspmigahidii	Khurais Road-150km beforeAl Ahsa	25° 11' 47" N 48° 19' 12" E	MG664283	MG664314
19.	D13	*T. hamiensis*var. *mandavillei*	Khurais - Al Ahsa road	25° 13' 55" N 48° 36' 16" E	MG664299	MG664330
20.	D16	*T. hamiensis*var. *qatarensis*	Al Ahsa - Qatar road	25° 16' 29" N 49° 41' 07" E	MG664295	MG664326
21.	D18	*T. hamiensis*var. *hamiensis*	Al Ahsa - Qatar road	25° 16' 30" N 49° 41' 09" E	MG664291	MG664322
22.	D19	*T. hamiensis*var. *hamiensis*	Al Ahsa - Qatar road - 25 km beforeSalwa	24° 49' 54" N 50° 40' 25" E	MG664292	MG664323
23.	D20	*T. hamiensis*var. *qatarensis*	Al Ahsa - Qatar road	24° 48' 40" N 50° 44' 26" E	MG664296	MG664327
24.	D21	*T. hamiensis*var. *qatarensis*	Al Ahsa - Qatar road	24° 48' 40" N 50° 44' 26" E	MG664297	MG664328
25.	D22	*T. hamiensis*var. *qatarensis*	Al Ahsa - Qatar road	24° 48' 40" N 50° 44' 26" E	MG664298	MG664329
26.	D24	*T. hamiensis*var. *hamiensis*	Al Ahsa - Qatar road -10 km before Alaudaidah	24° 27' 32" N 51° 02' 52" E	MG664293	MG664324
27.	D25	*T. hamiensis*var. *mandavillei*	Al Ahsa - Qatar road -10 km before Alaudaidah	24° 27' 32" N 51° 02' 52" E	MG664300	MG664331
28.	D27	*T. propinqua*sspmigahidii	Al Ahsa - Dammam road	25° 37' 33" N 49° 32' 12" E	MG664284	MG664315
29.	D28	*T. hamiensis*var. *hamiensis*	Al Ahsa - Dammam road	25° 37' 33" N 49° 31' 11" E	MG664294	MG664325
30.	D29	*T. propinqua*sspmigahidii	Shedgum-next tothe cement factory-Al Ahsa-Dammam road	25° 40' 07" N 49° 30' 31" E	MG664285	MG664316
31.	D30	*T. hamiensis*var. *mandavillei*	Shedgum-next tothe cement factory-Al Ahsa-Dammam road	25° 40' 07" N 49° 30' 31" E	MG664301	MG664332

the two regions (*rbc*L and *trn*L-F) from *Fagonia*, the most closely related genus to *Tetraena* and *Zygophyllum*, were downloaded from GenBank to use as the out-group (Table 3). Out-group choice was based on previous work on the genus *Zygophyllum* (Bellstedt *et al.*, 2008) and work on the sisters' genera to *Tetraena* and *Zygophyllum*, which is *Fagonia*.

Table 2. Herbarium specimens used in the present study for phylogenetic analyses.

No.	Taxa	Collection number	Collector's name	Date	Country	Herbarium
1.	*Tetraena hamiensis* var. *hamiensis* E4	M. 8153	Miller *et al.*	13/2/1989	Yemen	E
2.	*T. hamiensis* var. *hamiensis* E10	MTA 155	Abdullah M.	9/5/2012	Kuwait	E
3.	*T. hamiensis* var. *qatarensis* E9	21/2	Munton	21/1/1985	Oman	E
4.	*T. hamiensis* var. *qatarensis* K8	2	Vujo, K. J.	4/ 1979	Bahrain	K
5.	*T. hamiensis* var. *qatarensis* K9	10953	Boules, L.	29/3/1977	Qatar	K
6.	*T. propinqua* ssp. *migahidii* E6	6731	S. Collenette	27/4/1988	Saudi Arabia	E

Table 3. Sequences obtained from GenBank and previously used in the analysis of *Tetraena* and *Zygophyllum* plants (After Bellstedt *et al.*, 2008).

Taxa	GenBank accession for *rbc*L	GenBank accession for *trn*L-F
Fagonia cretica L. (out group)	AJ133855	AJ387942
F. indica Burm.f. (out group)	Y15018	AJ387943
F. luntii Baker (out group)	AJ133856	AJ387944
Tetraena mongolica Maxim.	Y15027	AJ387959
Zygophyllum album L.f.	AJ133861	AJ387963
Z. coccineum L.	AJ133863	AJ387965
Z. decumbens Delile	AJ133865	AJ387967
Z. decumbens Delile var. *decumbens*	EF655991	EF 656011
Z. fabago L.	Y15030	AJ387968
Z. sessilifolium L.	EF655997	EF656047
Z. simplex L.	EF655984	EF 656004
Z. simplex L.	Y15031	AJ387974
Z. xanthoxylum Engl.	AJ133872	AJ387975

DNA extraction

Leaf material from field-collected plants and herbarium specimens (Tables 1 & 2) were used for DNA extraction. Leaves were dried and stored in small polythene bags at -20°C. Total genomic DNA was extracted using the DNeasy Plant Mini Kit (QIAGEN) following the manufacturer's protocol. The isolated DNA was stored at -20°C until further use.

Choice of molecular markers

The phylogenetic relationship of Saudi Arabian *Tetraena* and *Zygophyllum* taxa was clarified using two different chloroplast regions (*rbc*L and *trn*L-F regions) based on results from the previous work on *Tetraena* and *Zygophyllum* (Beier *et al.*, 2003; Bellstedt *et al.*, 2008).

DNA amplification

The DNA template amplified using PCR (Polymerase Chain Reaction). The PCR used different primers to amplify the *rbc*L and the *trn*L-F chloroplast DNA (cpDNA) regions. The PCR amplifications for each region were carried out in 25 µl reactions using 2 µl of template DNA, 12.5 µl 2x BioMix (Bioline), 2 µl of each primer [1-10 mM] and, 6.5 µl of distilled water. The *rbc*L gene was amplified using the forward primer 20bp at 1F (5'- ATGTCACCACAAACAG AAAC-3') and reverse primer 26bp at 1460R (5'- TCCTTTTAGTAAAAGATTGGGCCGAG-3') based on Savolainen *et al.* (2000a, b). The PCR conditions for the *rbc*L amplification used the protocol as outlined in Bellstedt *et al.* (2008), with some modifications for some accessions. The reaction condition was 5 min at 94°C, followed by 30 cycles of denaturation at 94°C for 30s, annealing temperature at 50-53°C for 50s, extension at 72°C for 60s, followed by a final extension for 6 min at 72°C. The *trn*L-F region was amplified using the forward primer 20 bp at c (5'-CGAAATCGGTAGACGCTACG-3') and reverse primer f (5'-ATTTGAACTGGTGACACGAG-3') based on Taberlet *et al.* (1991). The PCR conditions for the *trn*L-F amplifications were used the following program based on Bellstedt *et al.* (2008) which included 5 min at 94°C, followed by 35 cycles of denaturation at 94°C for 60s, annealing temperature at 55°C for 60s, extension at 72°C for 90s, followed by a final extension for 6 min at 72°C.

PCR product purification and sequences

PCR reactions used an automatic sequencer ABI3730XL (Macrogen Sequencing System, Korea) for purification and sequencing. For each sequence, the complementary bi-directional sequence strands were trimmed and assembled into a contig and manually edited using SeqMan software 6.1, Lasergene DNAStar 6.1 Windows 32 (DNAStar Corporation, Madison, WI, USA). All sequences were aligned automatically by BioEdit v.7.0.4.1 (Hall, 1999) or Clustal X (Thompson *et al.*, 1997) followed by extensive manual adjustments. The two alignments were combined in one matrix using MacClade v. 4.07 (Maddison and Maddison, 2003).

Phylogenetic analyses

Maximum Parsimony: Separate analyses of *rbc*L and *trn*L-F data, and of combined chloroplast (*rbc*L and *trn*L-F) data were performed to infer relationships of Saudi Arabian taxa of *Tetraena* and *Zygophyllum* using the Maximum Parsimony approach, implemented with the computer program PAUP* 4.06 b10 for 32-bit Microsoft Windows XP (Swofford, 2001). Bootstrap support analysis (Felsenstein, 1985; Felsenstein and Kashino, 1993) was implemented in PAUP* 4.06 (Swofford, 2001) to estimate the support value of individual and combined data sets with 1000 pseudoreplicates of the data using the heuristic search strategy.

Bayesian analysis

The *rbc*L and *trn*L-F and combined chloroplast (*rbc*L and *trn*L-F) were analysed to infer relationships of Saudi Arabian *Tetraena* and *Zygophyllum* plants using Bayesian inference (Mau *et al.*, 1999; Rannala and Yang, 1996) of the separate and combined data. Bayesian analysis used the Markov Chain Monte Carlo (mcmc) simulation programme, MrBayes version 3.1.2 (Huelsenbeck and Ronquist, 2001; Ronquist and Huelsenbeck, 2003). The best fit model of molecular evolution for each individual and combined data set was selected using the Akaike Information Criterion (AIC), calculated with MrModeltest 2.2 (Nylander, 2004). The general time reversible model with gamma and proportion of invariable sites of (GTR+I+G) was selected for all partitions as the best fit model. Five million generations were performed and 5000 trees were saved (sampling one tree per 1000 generations). Runs were repeated twice to confirm results, and typically 0.25% (c. 1250 trees) of the samples were discarded as burn-in. Majority rule consensus trees were constructed from the remaining trees to obtain posterior probabilities using PAUP* programme.

Results and Discussion

Parsimony analyses

The characteristics obtained by Parsimony Analyses of the individual and combined datasets for the taxa are summarizes in Table 4.

The *trn*L-F Parsimony analysis of 44 sequences yielded 100 of most parsimonious trees. All trees were saved and the strict consensus was generated (not shown). The *rbc*L Parsimony analysis of 40 sequences yielded 100 of the most parsimonious trees. All trees were saved and the strict consensus was generated (not shown). In case of combined cpDNA, the aligned matrix of combined chloroplast (*rbc*L and *trn*L-F) sequences was 2505 bp in length. Parsimony analysis of 44 sequences produced 100 of the most parsimonious trees. All trees were saved and the strict consensus was generated (Fig. 1).

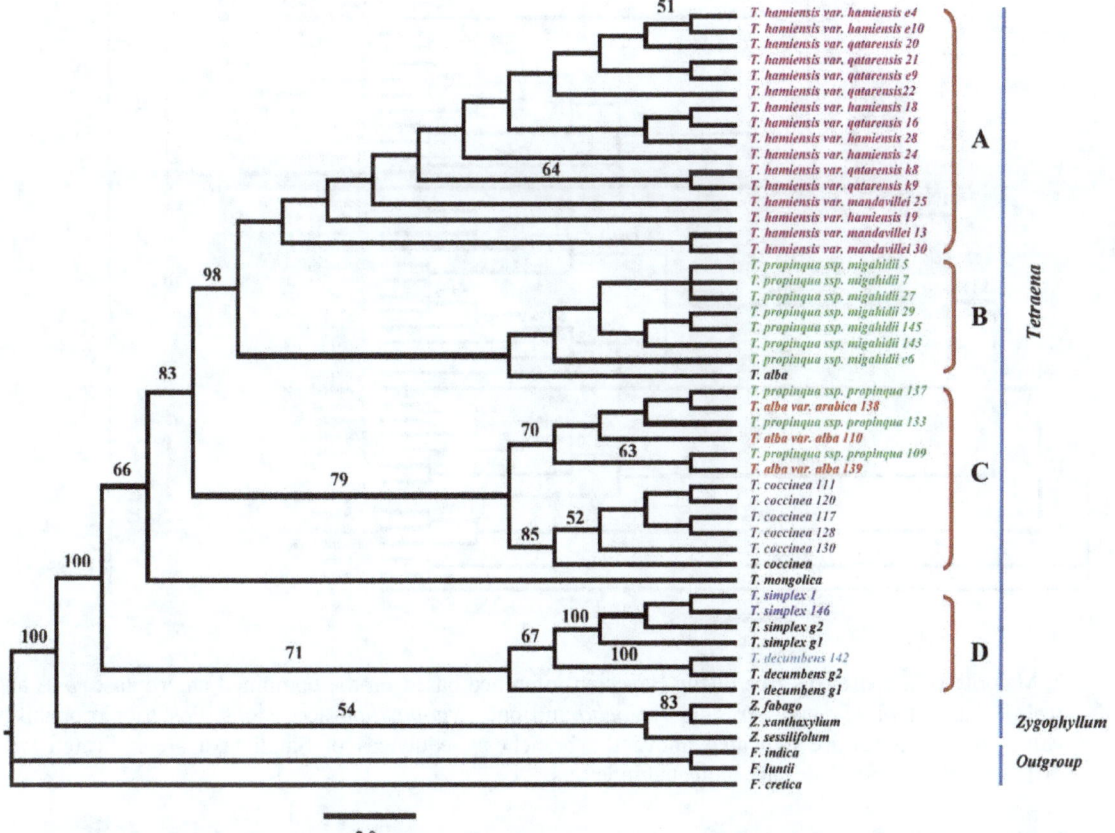

Fig. 1. One of 100 most equally parsimonious trees from analysis of the combined chloroplast of *rbc*L and *trn*L-F data set, using maximum parsimony for 43 Saudi Arabian *Tetraena* and one *Zygophyllum* accessions. Numbers above nodes are bootstrap (BS) support percentage values for clades supported above a 50% bootstrap value from 100000 replicates. Sequences of Saudi taxa are indicated with different colours and clades are indicated in letters.

Bayesian analyses

The best fitting model retrieved by MrModeltest as the most likely evolutionary model for all individual and combined data sets was the GTR+I+G model. Majority rule consensus trees were derived from 5000 trees from each analysis of the separate trnL-F (not shown) and rbcL (not shown) partitions and from combined chloroplast (Fig. 2) data sets. Burn-in was reached after 1250 generations for all partitions and for the combined matrix.

The represent study represents the first molecular phylogenetic study of the genus *Tetraena* and *Zygophyllum* in Saudi Arabia. Maximum Parsimony analysis and Bayesian criteria of the individuals and combined dataset of the *rbc*L and the *trn*L-F chloroplast DNA sequences used to study the phylogenetic relationships of *Tetraena* and *Zygophyllum* taxa in Saudi Arabia. The most notable similarity with respect to the individual and combined analysis regarding the overall topologies of the Maximum Parsimony and Bayesian trees are quite similar.

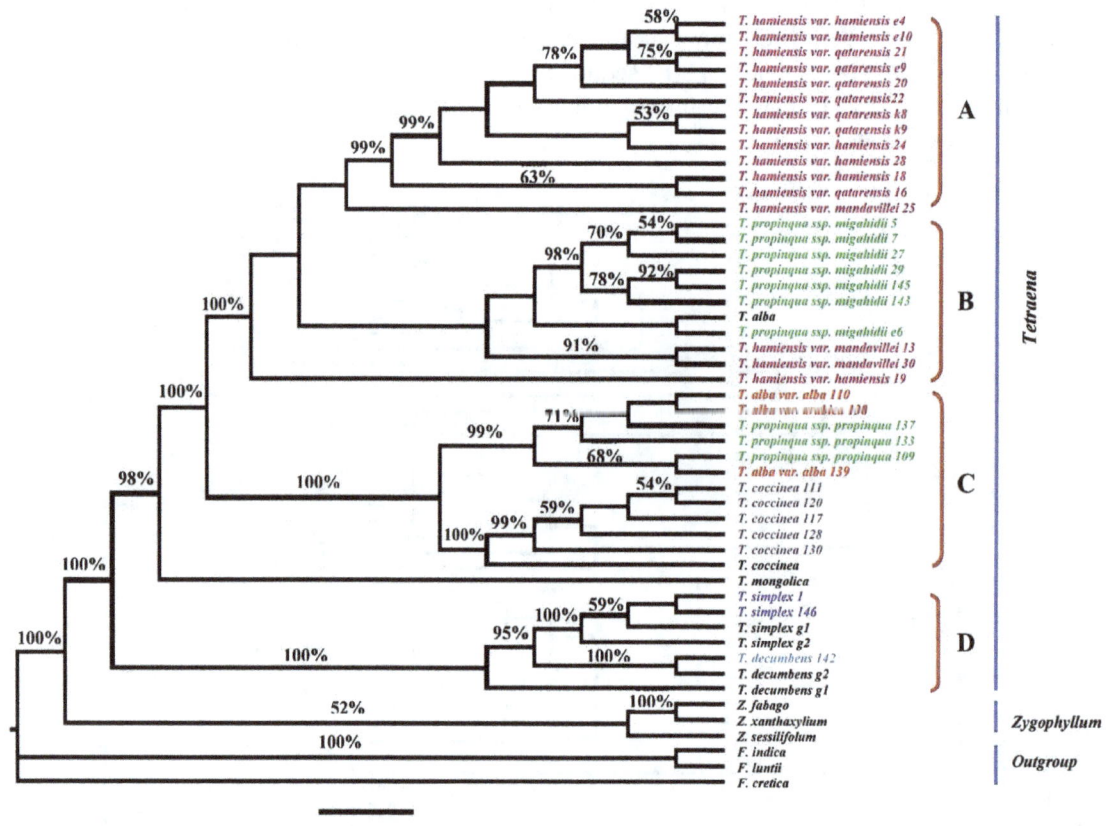

Fig. 2. Majority-rule consensus tree of the Bayesian inference based on the combined chloroplast *rbc*L and *trn*L-F data set of 43 Saudi Arabian *Tetraena* and one *Zygophyllum* accessions. Posterior probability values of the nodes are indicated above the branches. Sequences of Saudi taxa are indicated with different colours and clades are indicated in letters

Each of the *Tetraena* and the *Zygophyllum* genera appear as a monophyletic group with strong support in all phylogenies. In all phylogenetic analysis, the sequences of *T. mongolica* Maxim. (the type species of *Tetraena*), downloaded from GenBank, were nested within the rest of Saudi Arabian *Tetraena*. This finding agrees with Sheahan and Chase (2000) and supports the new classification of Beier *et al.* (2003). *Z. fabago* and *Z. xanthoxylum* (Asian species) samples that were downloaded from GenBank are nested together in one clade as monophyletic group in all phylogenies of the Maximum Parsimony and Bayesian analysis (Figs 1 & 2). Molecular phylogenetic results of the *rbc*L and *trn*L-F individually or in combination datasets analysis in this study suggested that *Z. fabago* species is differing from other samples. Moreover, the strong agreement of the findings of the morphological studies (Alzahrani, 2017; Alzahrani and Albokhari, 2017a, b;) and molecular phylogenetic analysis in this study support the classification of Beier *et al.* (2003) to separate *Tetraena* and *Zygophyllum* plants into two genera. Molecular

Table 4. Characteristics of the individual and combined datasets from Parsimony analysis.

Phylogenetic information	*rbc*L	*trn*L	Combined cpDNA
Number of accession	46	50	50
Aligned length	1434	1071	2505
No. of constant characters	1286	792	2078
No. of variable characters	56	134	190
No. of informative characters	92	145	237
No. of most equally maximum Parsimony trees	100	100	100
Length of shortest trees (steps)	194	457	673
Consistency index (CI)	0.8144	0.7287	0.7296
Retention index (RI)	0.9032	0.8041	0.8189
Rescaled consistency index (RC)	0.7356	0.5859	0.5974

phylogenetic of the cpDNA analysis divided Saudi Arabian *Tetraena* plants into six groups: *T. hamiensis* (Schweinf.) Beier & Thulin, *T. propinqua* (Decne.) Ghazanfar & Osborne, *T. alba* (L. f.) Beier & Thulin, *T. coccinea*, *T. simplex* (L. f.) Beier & Thulin, and *T. decumbens* (Delile) Beier & Thulin.

Acknowledgements

We are grateful to the curators and members of the staff of the herbaria K and E for allowing us to study and borrow specimens. Thanks are due to King Abdulaziz University, Jeddah, Saudi Arabia for providing research facilities

References

Al-Arjany, K.M. 2011. Molecular taxonomic perspective and eco-physiological variations of some species of *Tribulus, Zygophyllum* and *Fagonia* genera of family Zygophyllaceae in Saudi Arabia. Master dissertation, King Saud University, Saudi Arabia.

Alzahrani, D.A. 2017. Systematic studies on the Zygophyllaceae of Saudi Arabia: Two new subspecies combination in *Tetraena* Maxim. Saudi J. Biol. Sci. DOI: 10.1016/j.sjbs.2016.12.022.

Alzahrani, D.A., and Albokhari, E.J. 2017a. Systematic studies on the Zygophyllaceae of Saudi Arabia: a new variety and new variety combination in *Tetraena*. Saudi J. Biol. Sci. **24**: 1574–1579.

Alzahrani, D.A. and Albokhari, E.J. 2017b. Systematic studies on the Zygophyllaceae of Saudi Arabia: new combinations in *Tetraena* Maxim. Turk. J. Bot. **41**: 96–106.

Azevedo, L.B. 2014. Development and application of stressor-response relationships of nutrients. Ph.D. Thesis, Radboud University Nijmegen, the Netherlands.

Beier, B.A., Chase, M.W. and Thulin, M. 2003. Phylogenetic relationships and taxonomy of subfamily Zygophylloideae (Zygophyllaceae) based on molecular and morphological data. Plant Syst. Evol. **240**: 11–39.

Bellstedt, D.U., Van Zyl, L., Marais, E.M., Bytebier, B., de Villiers, C.A., Makwarela, A.M. and Dreyer, L.L. 2008. Phylogenetic relationships, character evolution and biogeography of southern African members of *Zygophyllum* (Zygophyllaceae) based on three plastid regions. Mol. Phylogenet. Evol. **47**: 932-949.

Boulos, L. 1978. Materials for a Flora of Qatar. Webbia **32**: 369–396.

El-Hadidi, M.N. 1977. Two new *Zygophyllum* species from Arabia. Publications from Cairo University Herbarium **7&8**: 327-329.

El-Hadidi, M.N. 1980. On the taxonomy of *Zygophyllum* section *Bipartita*. Kew Bull. **35**: 335–340.

Engler, A. 1931. Zygophyllaceae. *In*: Engler A., Prantl K. (2nd ed.) *Die Naturlichen Pflanzenfamilien* **19**: 144-184. Engelmann, Leipzig.

Felsenstein, J. 1985. Confidence limits on phylogenies: an approach using the bootstrap. Evolution **39**: 783–791.

Felsenstein, J. and Kashino, H. 1993. Is there something wrong with the bootstrap on phylogenies? A reply to Hillis and Bull. Syst. Biol. **42**: 193–199.

Ghazanfar, S.A. and Osborn, J. 2015. Typification of *Zygophyllum propinquum* Decne. and *Z. coccineum* L. (Zygophyllaceae) and a key to *Tetraena* in SW Asia. Kew Bull. **70**: 38.

Hall, T.A. 1999. BioEdit, a user-friendly biological sequence alignment editor and analysis program for Windows. Nucleic Acids Symp. Ser. **41**: 95–98.

Hammad, I. and Qari, S.H. 2010. Genetic diversity among *Zygophyllum* (Zygophyllaceae) populations based on RAPD analysis. Genet. Mol. Res. **9**: 2412–2420.

Hosny, A.I. 1988. Genus *Zygophyllum* L. in Arabia. Taeckholmia **11**: 19–32.

Huelsenbeck, J.P. and Ronquist, F. 2001. MrBayes: Bayesian inference in phylogenetic trees. Bioinformatics **17**: 754–755.

Hussein, S.R., Kawashty, S.A., Tantawy, M.E. and Saleh, N.A. 2009. Chemosystematic studies of *Nitraria retusa* and selected taxa of Zygophyllaceae in Egypt. Plant Syst. Evol. **277**: 251–264.

Louhaichi, M., Salkini, A.K., Estita, H.E. and Belkhir, S. 2011. Initial assessment of medicinal plants across the Libyan Mediterranean coast. Adv. Environ. Biol. **5**: 359-370.

Ma, Y. and Zhang, S. 1990. Study on the systematic position of *Tetraena*. Acta. Phytotax. Sin. **28**: 89–95.

Maddison, D.R. and Maddison, W.P. 2003. MacClade V. 4.07: Analysis of phylogeny and character evolution. Sinauer Associates, Sunderland, MA.

Mau, B., Newton, M.A. and Larget, B. 1999. Bayesian phylogenetic inference via Markovchain Monte Carlo methods. Biometrics **55**: 1–12.

Mosti, S., Raffaelli, M. and Tardelli, M. 2012. Contribution to the Flora of Central-Southern Dhofar (Sultanate of Oman). Webbia **67**: 65–91.

Norton, J., Abdul Majid, S., Allan, D., AlSafran, M., Böer, B. and Richer, R. 2009. An Illustrated Checklist of the Flora of Qatar. Browndown Publications, Gosport, UK. 67 pp.

Nylander, J. 2004. MrModeltest Q. Uppsala: Evolutionary Biology Centre, Uppsala University.

Rannala, B. and Yang, Z. 1996. Probability distribution of molecular evolutionary trees: a new method of phylogenetic inference. J. Mol. Evol. **43**: 304–311.

Ronquist, F. and Huelsenbeck, J.P. 2003 MrBayes 3: Bayesian phylogenetic inference under mixed models. Bioinformatics **19**: 1572–1574.

Sakkir, S., Kabshawi, M. and Mehairbi, M. 2012. Medicinal plants diversity and their conservation status in the United Arab Emirates (UAE). J. Med. Plants Res. **6**: 1304–1322.

Savolainen, V., Fay, M.F., Albachi, D.C., Backlund, A., Van der Bank, M., Cameron, K.M., Johnson, S.A., Lledo, M.D., Pintaud, J.C., Powell, M., Sheahan, M.C., Soltis, D.E., Soltis, P.S., Weston, P., Whitten, W.M., Wurdack, K.J. and Chase, M.W. 2000a. Phylogeny of the eudicots: a nearly complete familial analysis based on *rbc*L gene sequences. Kew Bull. **55**: 257–309.

Savolainen, V., Chase, M.W., Hoot, S.B., Morton, C.M., Soltis, D.E., Bayer, C., Fay, M.F., De Bruijn, A.Y., Sullivan, S. and Qiu, Y. 2000b. Phylogenetics of flowering plants based on combined analysis of plastid *atp*B and *rbc*L gene sequences. Syst. Biol. **49**: 306–362.

Sheahan, M.C. and Chase, M.W. 1996. A phylogenetic analysis of Zygophyllaceae based on morphological, anatomical and *rbc*L DNA sequence data. Bot. J. Linn. Soc. **122**: 279–300.

Sheahan, M.C. and Chase, M.W. 2000. Phylogenetic relationships within Zygophyllaceae based on DNA sequences of three plastid regions, with special emphasis on Zygophylloideae. Syst. Bot. **25**: 371–384.

Swofford, D. 2001. PAUP*: Phylogenetic analysis using parsimony (*and other methods), version 4.06 b10. Sinauer, Sunderland, Massachusetts, USA.

Taberlet, P., Gielly, L., Pautou, G. and Bouvet, J. 1991. Universal primers for amplification of three non-coding regions of chloroplast DNA. Plant Mol. Biol. **17**: 1105–1109.

Takhtajan, A.L. 1987. Flowering Plant. Komarov Botanical Institute, Russia.

Thompson, J.D., Gibson, T.J., Plewniak, F., Jeanmougin, F. and Higgins, D.G. 1997. The CLUSTAL_X windows interface, flexible strategies for multiple sequence alignment aided by quality analysis tools. Nucleic Acids Res. **25**: 4876–4882.

Thulin, M. 1993. Zygophyllaceae *In*: Thulin, M. (Ed.) Flora of Somalia. Royal Botanical Gardens, Kew **1**: 176–189.

Van Huyssteen, D.C. 1937 Morphologisch-systematische studien über die gattung *Zygophyllum*. Dissertation. Berlin.

Van Zyl, L. 2000. A systematic revision of *Zygophyllum* in the southern African region. Ph.D. thesis, University of Stellenbosch, Stellenbosch.

14

MORPHOLOGY AND MOLECULAR PHYLOGENY OF THE MARINE DIATOM *NITZSCHIA DENTATUM* SP. NOV. AND *N. JOHORENSIS* SP. NOV. (BACILLARIOPHYCEAE) FROM MALAYSIA

S.N.P. SURIYANTI[1] AND G. USUP

School of Environmental Science and Natural Resources, Faculty Science and Technology, Universiti Kebangsaan Malaysia, 43600 Bangi, Selangor, Malaysia

Keywords: Girdle; *Hantzschiod*; Indented valves; Jagged; New species; Pennate diatom

Abstract

The marine diatom *Nitzschia dentatum* sp. nov. isolated from seawater samples of Kudat and *N. johorensis* sp. nov. isolated from beach sand samples of Sibu Island, Malaysia, have been described in this paper. Morphological identification, molecular phylogeny and toxin analyses were executed on the pure non-axenic algal cultures designated as KD89 and PS8, respectively. The main distinguishing feature of *N. dentatum* sp. nov. compared to other species is the jaggedcingulum structure which is only unique to this species. Meanwhile, *N. johorensis* sp. nov.is strongly characterized by the '*hantzschioid*'and '*nitzschioid*' symmetry dimorphisms; a common diagnostic feature but rarely described in other *Nitzschia* species. Identification of both strains was made based on the frustule diagnostic features and verified using the partial large ribosomal subunit DNA sequences. The results have confirmed that these two speciesare independent entities and novel species that have not been documented elsewhere. A notable finding from the Maximum Likelihood (ML), Maximum Parsimony (MP) and Bayesian Index (BI) analyses have also revealed that *Nitzschia* species that have indentation in the middle of valves have been consistently grouped as same clade with high bootstrap values. The extracts of both species did not show detectable amount of domoicacid and have therefore, been classified as non-toxic. This discovery contributes to the documentation of *Nitzschia* species worldwide.

Introduction

Nitzschia Hassall is represented by 1,405 diatom species worldwide comprising the freshwater, brackish water and marine environments, with only half of it been accepted taxonomically (Guiry and Guiry, 2017). The species of *Nitzschia* are ecologically important as bioindicator (Maznah and Mansor, 2002; Trobajo*et al*., 2004; Trobajo *et al*., 2009), endosymbiont (Lee, 2011) and also as aquaculture live feed (Chu *et al*., 1996). Some *Nitzschia* species from the tropics regions are toxic. The first discovery of toxic *Nitzschia* species was identified from prawn pond samples in Vietnam (Lundholm and Moestrup, 2000) while others have been collected from estuarine sites such as in Malaysia (Suriyanti and Gires, 2015)and lagoon samples from the Southwest Mediterranean Sea (Smida *et al*., 2014). There has been no report of harmful blooms to date that were associated with *Nitzschia*.

In Malaysia, the distribution of this genus has been listed along with other diatoms during field surveys and studies (Cleve, 1901; Nather-Khan, 1990; Shamsudin, 1990; Aishah and Nooraida, 1994; Aishah, 2005; Fareha *et al*., 2011; Saifullah *et al*., 2014). Taxonomical

---

[1]UTM Ocean Thermal Energy Centre (OTEC), Ground Floor, Block Q, Universiti Teknologi Malaysia, Jalan Sultan Yahya Petra, 54100 Kuala Lumpur, Malaysia.
Corresponding author. Email: sue_0586@yahoo.com.my

identification of the diatom *Nitzschia* has been highly dependent on discernable valve characters under light microscopy and the molecular data on the existing *Nitzschia* species was lacking. In molecular phylogeny of microalgae, the D1–D3 domain of LSU rDNA has been commonly used and proven as suitable marker for species delineation (Ki and Han, 2005; Sonnenberg *et al.*, 2007; Lundholm *et al.*, 2002). Currently, 14 species of *Nitzschia* have been compiled from Malaysian waters (Suriyanti, 2017). In the present research, two species of *Nitzschia* have been identified based on morphological and molecular characteristics and hitherto reported as new to science. This research is part of the results obtained from *Nitzschia* distribution study in Malaysia (Suriyanti, 2017).

Materials and Methods

Nitzschia cell isolate KD89 was obtained from marine net haul sample of Kudat, Borneo and isolateNPS8 from sand sediment sample of Sibu Island, Johor, Malaysia. Both isolates were established into pure non-axenic clonal cultures. All cultures were grown in silica-enriched media modified from SWII (Iwasaki, 1961) concoction adjusted to 30 Practical Salinity Unit (PSU)and maintained at 26 °C under 12:12 hour light : dark photo cycle.

The removal of organic matter by acid treatment was done according to Renberg (1990).Cultured cells were harvested by centrifugation at 8000 revolutions per minute (RPM) for 10 minutes and the supernatant was discarded. Hydrogen peroxide (30%) was added to the cell pellet and heated at 85°C for two to three hours. After oxidization, hydrogen peroxide was discarded and the samples were treated with 10% hydrochloric acid for several days at room temperature. After treatment, cells were rinsed two to three times with distilled water and stored in 70% ethanol.

The average size of valves was obtained by measuring the specimens from the first batch of clonal cultures to minimize size reduction due to mitotic division. Cleaned diatom valves were mounted on a glass slides using Naphrax mountant (Brunel Microscope Ltd., U.K.) and viewed under a light microscope (Olympus BX51TF, Japan) equipped with built-in camera (Olympus U-TV1x, Japan)at 20 × magnification. A minimum of 30 cells were randomly selected for length and width measurements by using Analy SIS Life Science Professional software version 3.0 (Build 1243).

The ultra-structural valve characteristics for species identification were observed by using electron microscopy. For scanning electron microscopy (Leo 1450 VP, United Kingdom), cleaned specimens were dried overnight on cover slips and mounted on a stub for gold-palladium coating before viewing. For transmission electron microscopy (Philips CM12, Netherland), cleaned specimens were mounted on formvar-coated copper grids. The diagnostic features for identification were the valve outlines, internal valves, valve striations, eccentricity of the raphe system and the presence of poroids in the raphe canals.

Molecular data were obtained from DNA samples of fresh specimens of pure clonal cultures. Diatom cells from the clonal cultures were harvested by centrifugation at 8000 rpm for 10 minutes. The genomic DNA was extracted using Gene JET Plant Genomic DNA Purification Mini Kit. Targeted LSU rRNA gene was amplified using primer set D1R 5'-ACC CGC TGA ATT TAA GCA TA-3' (Scholin *et al.*, 1994) and D3B 5'-TCG GAG GGA ACC AGC TAC TA-3' (Nunn *et al.*, 1996). Total PCR reaction volume of 50µl contained 0.2 g/µl bovine serum albumin, 0.2 mMdNTPs, 0.5 µM of forward and reverse primers, 1x Taq Buffer, 1.5 mM $MgCl_2$, 1.25 U Taq Polymerase (Fermentas #EP0402) and DNA template. The amplification condition was set atone initial denaturation at 94°C for 2 min, 30 cycles of 94°C for 30s, 60°C for 30s, 72°C for 30s and followed by final extension at 72°C for 2min (Lundholm *et al.*, 2002). Amplified products

116

were visualized by electrophoresis on a 1% agarose gel pre-casted with Red Safe nucleic acid staining solution (iNtRON Biotechnology Cat. No.21141). Purified products were sent to First Base Laboratories (Malaysia) for sequencing using the same primer set.

In the analysis of multiple sequence alignment, the quality of DNA sequences were checked manually using Bioedit version 7.0.9.0 (Hall, 1999). Reverse sequences were reverse-complemented with the forward sequences using optimal GLOBAL pairwise alignment. Trimmed sequences were saved in FASTA file and uploaded to NCBI online database for query using BlastX (Zhang *et al.*, 2000). Other *Nitzschia* LSU sequences were downloaded from the http://www.ncbi.nlm.nih.gov/ websiteas well. Multiple sequence alignment was executed using Muscle in MEGA Version 6 (Tamura *et al.*, 2013) and Clustal X version 1.81 (Thompson *et al.*, 1997) and saved in NEXUS format.

Phylogenetic analyses were performed using ML and MP algorithm in PAUP* Version 4.0b10 (Swofford, 2003). For ML analysis, the best model GTR+G+I was generated using Modeltest3.7 (Posada and Crandall, 1998). Heuristic search was used in the MP analysis. Tree reliability was estimated using bootstrap method with 1000 replicates of data set for ML and MP. Bayesian Analysis was used to generate the best phylogenetic tree using prior probability Monte Carlo Markov Chains (MCMC) Method with 490000 generation in Mr Bayes version 3.1.2 (Ronquist and Huelsenbeck, 2003). Distance Analysis was generated using PAUP* Version 4.0b10 (Swofford, 2003).

Results and Discussion

Nitzschia dentatum Suriyanti S.N.P. & G.Usup, **sp. nov.** (Figs 1A–L).

Diagnosis: The outline and valve characteristics of *N. dentatum* sp. nov. is compatible to species categorized under the section of *Lanceolatae*. Translated from Cleve and Grunow 1830 (as cited in Mann, 1978), the section *Lanceolatae* is defined as "lanceolate-linear or rarely oval, highly eccentric keels, keel puncta not prolonged". Many species in this section have cell dimensions close to *N. dentatum* sp. nov. Due to the high variability of the sizes, other features such as the shape of valve and presence of central interspace were used in combination to select the most proximate species from its allies. *Nitzschia dentatum* sp. nov.has slender and narrow cell outlines, most similar with *N. inconspicua* Grunow, *N. frustulum* (Kutzing) Grunow and *N. pusilla* Grunowin this section (Table 1). *N. dentatum* sp. nov.lack the central interspace, different from *N. frustulum* and *N. pusilla* except *N. inconspicua*. *N. dentatum* sp. nov. differs from *N. inconspicua* and the rest of it allies by its relatively high density of striae (78 in 10 μm); whereby other species only have average maximum number of 30–40 striae in 10μm. The main distinguish features that are only present in this species is the cingulum structure that is jagged which resembles the 'teeth' which could be easily distinguished from the cingulum of other species of *Nitzschia*. This feature is similar with the lateral extensions of closed copula in *Rhabdonema* sp. (Round *et al.*, 1990), with the exception that *N. dentatum* sp. nov. has an open-type girdle bands.

Type: MALAYSIA. Kudat, Sabah, seawater sample, 6° 51' N, 116° 51' E, collected 8 December 2013, isolated by capillary washing technique on 10 December 2013, Suriyanti and Usup.

Holotype: Voucher # KD89, deposited in the Marine Microbes and Biotechnology Laboratory, Universiti Kebangsaan Malaysia.

NCBI Accession no.: KX839243.

Notes: Marine habitat; cells solitary; each cell contains two yellow-brown chloroplasts; valves small and narrow; cell outline lanceolate and tapering towards the apices; rectangular in girdle view; length 17.0–18.0 μm, width 2.5–4.0 μm; apices are slightly capitate; wide pervalvar axis; raphe more or less eccentric; raphe continuous in external view; rectangular fibulae, not widely separated in the middle of valve; jagged structure of the girdle band was observed in each valve; fibulae on the diagonal side of valves (*nitzschioid*); terminal fissures are slightly hooked to the sameside; 13 fibulae in 10µm; terminal fissures bent towards the same side which end in a large *helictoglossa* internally; raphe ending at the centre only observable from internal view; it is raised in a simple shallow raphe canal which contains poroids; interstriae are raised externally, striae not interrupted by laternal sterna;78 striaein 10µm;single row of round to rectangular poroids, occluded by simple-type hymen perforation, 9–10 in 1 µm across; single strip of jagged cingulum structure with striae is observed for each valve; 1 row of poroid in the cingulum; jagged strip is lined by striaethat containporoids.

Etymology: Named after the unique jagged structure of the girdle band. 'Dentatum' is Latin for 'toothed'.

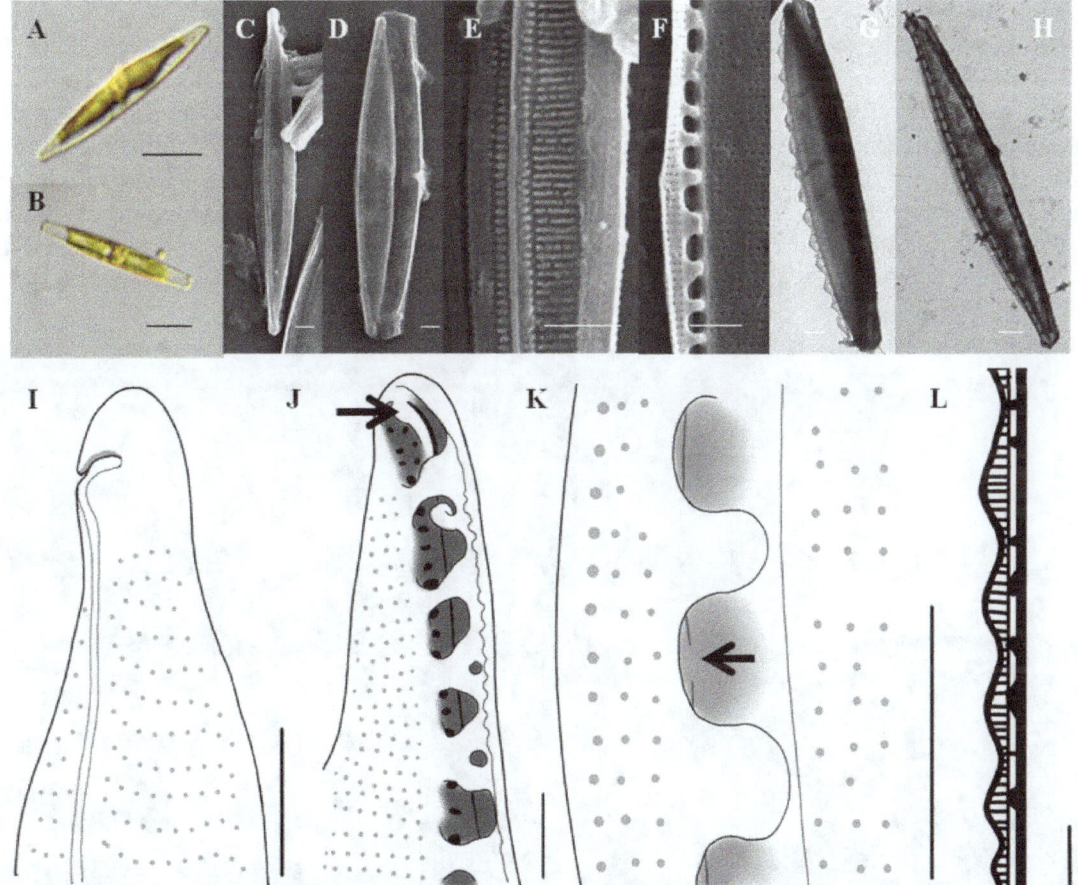

Figs 1A–L. LM, SEM, TEM and drawings of *N. dentatum* sp. nov. voucher #KD89. LM (A–B). A: Whole cell in valve view showing two yellow-brown chloroplasts; B: Rectangular in girdle view. (Scale bar = 10 µm) SEM (C–F). C: Valve view showing slightly capitate ends; D: Wide pervalvar axis of intact valves, showing the position of the jagged cingulum in girdle; E: Continuous raphe slit from the external view; F: arrangement of the fibulae, without large central interspace. TEM (G–H). G: Partially intact dental-like cincture at the margin of the valve; H: Fibulae on the diagonal side of the valves ('*nitzschioid*' symmetry). Scale bar = 1 µm. Illustration (I–L): I: Terminal fissure bent towards the same side in both ends; J: Terminal raphe ends in *helictoglossa* internally. K: central raphe ending observed from the internal view; L: close-up of the jagged structure of the girdle with horizontal striae. (Scale bar = 1 µm).

118

Nitzschia johorensis Suriyanti S.N.P. & G. Usup, **sp. nov**.   **(Figs 2A–I)**.

Diagnosis: *Nitzschia johorensis* sp. nov.is relatively small in size and best fit into the section *Lanceolatae*. After acid-treatment, the prominent girdle attachment of the epivalves and hypovalves remained intact, compared to other species whereby the frustules completely disintegrated after the same procedures of acid treatment. Observation of the valve symmetry in this species was made possible this way. Only small-sized *Nitzschia* species in this section were included for comparison. Overall, *N. johorensis* sp. nov. is most similar with *N. fonticola* except for habitat origin whereby the latter is strictly confined to the freshwater (Foged, 1971; Tudesque *et al.*, 2008), has a central interspace and only exist in *nitzschioid* form (Mann, 1978). On the other hand, the valve size of *N. johorensis* sp. nov.is compatible to the measurements of *N. tropica* Hustedt (Tudesque *et al.*, 2008) and *N. costei* Tudesque, Rimet *et.* Ector but differs in the shape of valve ends. In addition, *N. tropica* has widely separated fibulae in the middle (Tudesque *et al.*, 2008), a feature not present in *N. johorensis* sp. nov (Table 1). *N. johorensis* sp. nov. can be distinguished from *N. costei*by its thickened and imperforated siliceous marginal wall along the keel towards the apices, whereas *N. costei*has a double row of striae near the keel. The main feature in *N. johorensis* sp. nov. not found in other samples in this study is the *nitzschioid* and *hantzschioid* dimorphisms. It is rarely documented in other *Nitzschia* species (Mann, 1978; Round *et al.*, 1990), but it has been noted insome species such as the heterotrophic *N. alba* J.C. Lewin and R. A. Lewin (Lauritis *et al.*, 1967), the polar species *N. frigid* Grunow (Medlin and Hasle, 1990) and the middle-constricted *N. dubia* Smith (Mann, 1978).None of those species matched the description of *N. johorensis* sp. nov. morphologically and ecologically.

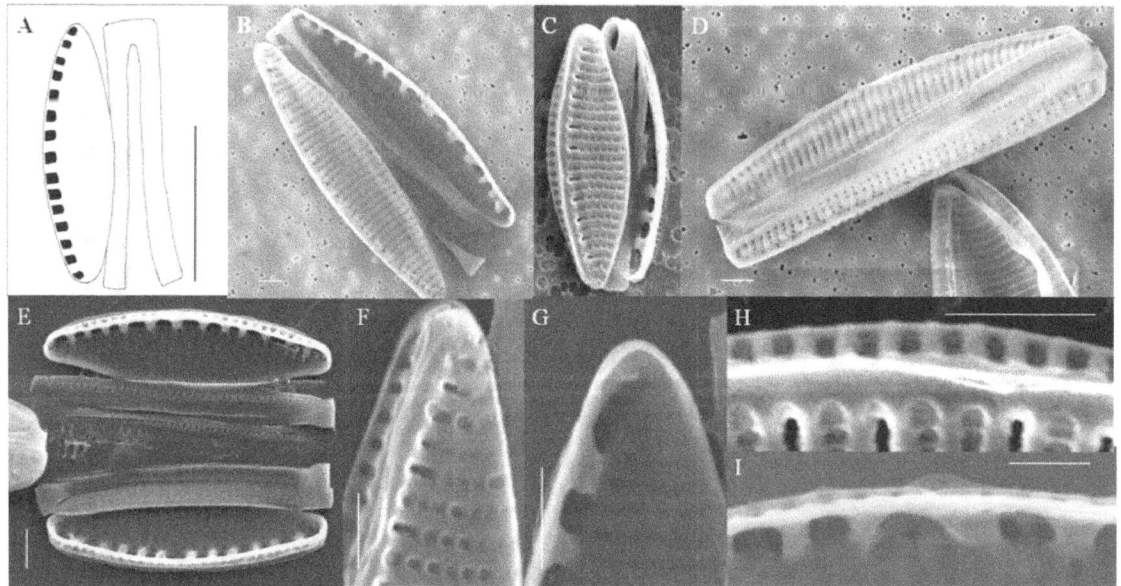

Figs 2A–I. Drawings and SEM micrographs and of *N. johorensis* sp. nov. isolate PS8. Illustration (A) A: Lanceolate valve outline. Scale bar = 10 μm. SEM (B–I). B: Irregular fibulae arrangement and width; C: Thick interstriae and margin, poroids more elongated near margin; D: Rectangular in girdle view of the intact valves; E: 'hantzschioid' symmetry of the raphe; closed-type of girdle band with single row of poroid; F: Terminal fissure, valve margin near apical ends without perforation; G: Terminal fissure ends in helictoglossa internally; H: Raphe is interrupted in the middle; I: raised raphe canal with pores. (Scale bar = 1μm).

Table 1. Morphometric data of species that have the closest similarity to *Nitzschia dentatum* sp. nov. and *N. johorensis* sp. nov. (n.d. = no data).

Species	Length (μm)	Width (μm)	Fibulae (10 μm)	Striae (10 μm)	Central interspace (+/−)	Reference
*N. costei*Tudesque, Rimet & Ector	8–45	2.5–4.5	9–12	23–27	+	Tudesque *et al.* (2008)
N. dentatum sp. nov. (n>30)	17.0–18.0	2.5–4.0	11–13	70–78	−	Present study
N. fonticola (Grunow) Grunow	10.0–55.0	2.5–4.5	n.d.	24–27	+	Kociolek (2011)
	13.5–22.0	4.0–5.0	9–11	22–26	n.d.	Foged (1971)
	7.0–46.0	2.5–5.5	10–12	26–30	+	Tudesque *et al.* (2008)
N. frustulum (Kutzing) Grunow	n.d.	n.d.	8–10	28–30	+	Mann (1978)
	10.8–34.0	3.0–3.9	13.3–15	26.6–30	+	Trobajo *et al.* (2013)
	12.0–14.0	3.0–4.0	14–15	45–52	+	Present study
N.inconspicua Grunow	4.1–15.3	2.3–3.1	8.9–17	23.7–30.4	+	Trobajo *et al.* (2013)
	12.1–14.4	2.0–3.2	3–4	28–31	−	Present study
*N. johorensis*sp. nov. (n>30)	7.1–11.8	1.8–3.5	11–12	30–33	−	Present study
N. pusilla Grunow	18.0–20.0	3.5–8.0	15–16	40–46	−	Coste and Ricard (1980)
	7.2–9.7	1.8–3.5	>9	>37	−	Present study
N. tropica Hustedt	14.5–44.6	3–3.7	8–10	23–25	+	Tudesque *et al.* (2008)
Nitzschia sp. 1 (PgMky44)	25.7–32.8	2.8–4.6	7–8	n.d.	+	Present study
*Nitzschia*sp.2 (KD90)	9.2–13.4	2.2–3.5	8–10	n.d.	−	Present study

Type: Malaysia. Pulau Sibu (Sibu Island), Johor, beach sand sediment, 2° 12' N, 104° 04' E, collected 5 June 2012, isolated by capillary washing technique on 7 June 2012, Suriyanti and Usup.

Holotype: Voucher no.PS8,deposited in the Marine Microbes and Biotechnology Laboratory, Universiti Kebangsaan Malaysia.

NCBI Accession no.: KX839235

Notes: Two chloroplasts at both ends; valve lanceolate, length 7.1–11.8 μm, width1.8–3.5 μm; slightly capitated ends; rectangular in girdle views; valve mantel is a one-row height of elongated poroids; fibulae are either on the same side (*hantschioid*) or on the diagonal side (*nitzschioid*) of the complimentary valve; fibulae coarse and wide; irregularly spaced; 11 in 10 μm; interstriae are raised externally, smooth on the internal surface; 33 striae in 10 μm; poroids are round to rectangular; more elongated towards the valve margins; terminal fissure hooked towards the valve face; valve near the raphe and apices thickened without perforations; *helictoglossa* ending internally; 1 row of poroids, 3–5 in 1 μm; 3–4 bands of semi-closed girdle type; single row of poroids in the copulae; raphe canal is raised with pores; central nodule in raphe slit; central interspace absent.

Etymology: The species epithet is named after the state (Johor) from where it was found.

Phylogenetic analyses: The LSU rDNA sequences used in the phylogenetic analyses were obtained from 11 *Nitzschia* species (Table 2) out of 14 total marine *Nitzschia* species recorded from Malaysia (Suriyanti, 2017). Amplification of the LSU rDNA region yielded product length of ca. 800 basepairs.   Taxa relationship of *Nitzschia* was inferred by the placement of *Pseudonitzschia americana* (Hasle) Fryxell and *Fragilariopsis kerguelensis* (O'Meara) Hustedt as out groups. These two genera were previously originated as two subsections in the genus *Nitzschia* beforeclassified as separate genera. The bootstraps (1000 replicates) were shown next to branch. The sequences used for the ML (Fig. 3), MP (Fig. 3), Bayesian Analysis (Fig. 4) and Distance Analysis (Table 3) were obtained from *Nitzschia* spp. recorded from Malaysia and retrieved from the Genbank BLAST query database. Those analyses have included 21 partial LSU nucleotide sequences including two out groups.

Table 2. *Nitzschia* culture strains used in the phylogenetic analyses.

Species	Location	Coordinate	Strain	GenBank
N. amabilis	Teluk Kumbar, Penang	5°17'4"U, 100°14'22''T	TK47	KX839238
Nitzschia sp. 1	Teluk Kumbar, Penang		PgMky44	KX839237
N. sigma	Kuala Selangor, Selangor	3°20'20"U, 101° 14'41"T	KS58	KX839241
N. lorenziana	Kuala Selangor, Selangor		KS55	KX839240
N. navis-varingica	Sungai Pendas, Johor	1°23'3"U, 103°37'30"T	P22C7	KX839243
N. johorensis sp. nov.	Pulau Sibu, Johor	2°13'34"U, 104°3'44"T	PS8	KX839235
N. pusilla	Pulau Tioman, Pahang	2°47'37"U, 104°12'7"T	TMN26	KX839236
N. dentatum sp. nov.	Kudat, Sabah	6°53'12"U, 116°49'31"T	KD89	KX839243
Nitzschia sp. 2	Kudat, Sabah		KD90	KX839242
N. frustulum	Kudat, Sabah		KD92	KX839245
N. inconspicua	Simpang Mengayau, Sabah	7°1'26"U, 116°44'34"T	TOB54	KX839239

All reconstructed phylogenetic trees showed almost identical topologies but differed in bootstrap values. The ML and MP phylogenetic tress were only distinguished in the placement of *Nitzschia* sp. 1.   Two distinct clades were generated in the trees in which *N. Soratensis*E. A. Morales *et* M. L. Vis and*N.* cf. *Fonticola* (Grunow) Grunow formed as sister clade to other *Nitzschia* spp. *N. dentatum* sp. nov. dan *N. johorensis* sp. nov. formed isolated branches on each phylogenetic tree and were supported by bootstrap 63 dan 57 in ML and 66 dan 61in MP, correspondingly. The prior probability in Bayesian inference was 0.9939 for*N. dentatum* sp. nov. and 0.716 for*N. johorensis* sp. nov.   The least genetic distance to delineate these species was estimated at 4.86% (Table 3). *N. dentatum* sp. nov. was grouped into the same clade as *N.* cf. *promare* Medlin, *N. pellucida* Grunow, *N. navis-varingica* N. Lundholm and Ø. Moestrup, *N. amabilis* Suzuki, *Nitzschia* sp. 1dan *N. lecointei* van Heurck. On the other hand, *N. johorensis* sp. nov. formed single branch and did not cluster with any other clades. Consistent grouping was observed in a clade comprising *N.* cf. *promare*, *N. pellucida*, *N. navis-varingica* and *N. amabilis* in all trees and supported by high bootstrap values (ML: 98, MP: 100, BI: 1).

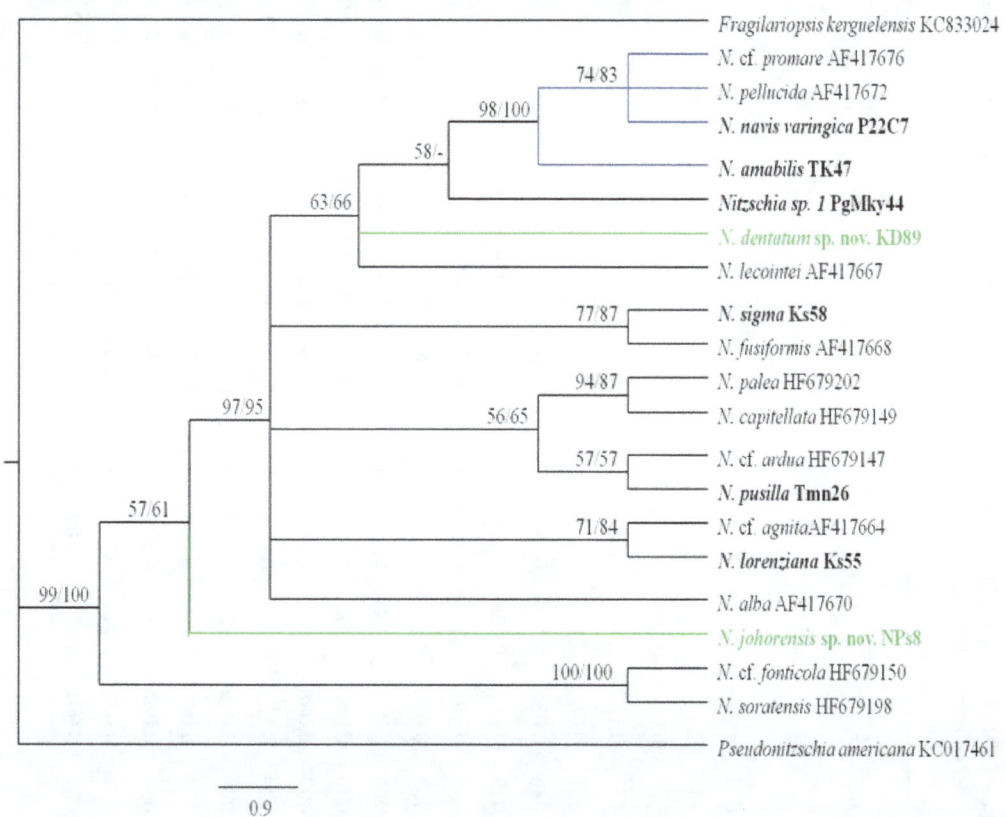

Fig. 3. Phylogenetic trees of *Nitzschia* spp. reconstructed based on the D1–D3 gene region of LSU rDNA using Maximum Likelihood and Maximum Parsimony with *P. americana* and *F. kergeulensis* as outgroups. Bold-lettered strains were obtained from this study (black) and proposed species (green). Species that are constricted in the middle valve grouped into the same clade consistently in all tress (blue). Bootstrap values with 50% majority are shown (ML/MP).

This study proposes two *Nitzschia* species into the section *Lanceolatae* based on morphological diagnoses and molecular evidence using LSU rDNA genes. As taxonomic conclusion should not solely rely on the valve characters, molecular characterisation based on the LSU rDNA regions was done to verify the phylogeny placement of the two new species. The D1–D3 region of LSU rDNA is a highly variable and a suitable marker for species identification (Ki and Han 2005; Sonnenberg *et al.*, 2007; Lundholm *et al.*, 2002) amongst the 12 more conserved D-domains. Highly conserved SSU marker on the other hand is less desirable in taxonomic purposes, but is helpful in depicting the original lineage of diatom (Zimmermann *et al.*, 2011; Smida *et al.,* 2014).

Phylogenetic tree reconstruction showed that the genus *Nitzschia* is not monophyletic, in agreement with its great variationin morphology. Based on the tree, *N. dentatum* sp. nov. is closely related to *N.* cf. *promare*, *N. pellucida*, *N. navis-varingica*, *N. amabilis*, *Nitzschia* sp. 1 and *N. lecointei*.None of these species has jagged girdle bands except for the proposed species *N. dentatum* sp. nov. Girdle character is one of the important features for species delineation (Mann, 1978; Round *et al.*, 1990; Lundholm and Moestrup, 2000). Furthermore, those species are genetically diverged at 7%–11%, respectively (Table 3).

Table 3. Genetic distance matrix of 21 *Nitzschia* spp. analyzed based on LSU rDNA partial gene sequences with *P. americana* and *F. kerguelensis* as outgroups. The distance was calculated using Kimura-two-model (Kimura, 1980).

	Species	1	2	3	4	5	6	7	8	9	10	11	12
1	AF417676 *Nitzschia* cf. *promare*	-											
2	*Fragilariopsis kerguelensis*	0.097	-										
3	KX839241 *N. sigma*	0.089	0.112	-									
4	AF417672 *N. pellucida*	0.022	0.105	0.095	-								
5	HF679150 *N.* cf. *fonticola*	0.109	0.107	0.131	0.117	-							
6	KC017461 *Pseudonitzschia americana*	0.092	0.027	0.098	0.095	0.099	-						
7	KX8392358 *N. johorensis*	0.110	0.090	0.127	0.120	0.123	0.090	-					
8	HF679202 *N. palea*	0.076	0.112	0.083	0.086	0.120	0.103	0.104	-				
9	AF417664 *N.* cf. *agnita*	0.080	0.111	0.086	0.086	0.118	0.096	0.104	0.067	-			
10	KX839243 *N. dentatum*	0.076	0.112	0.079	0.083	0.114	0.110	0.113	0.071	0.074	-		
11	HF679147 *N.* cf. *ardua*	0.085	0.120	0.085	0.095	0.111	0.114	0.097	0.055	0.085	0.082	-	
12	AF417668 *N. fusiformis*	0.082	0.105	0.063	0.082	0.117	0.096	0.109	0.083	0.073	0.080	0.078	-
13	KX839238 *N. amabilis*	0.034	0.107	0.085	0.038	0.109	0.101	0.116	0.074	0.072	0.073	0.092	0.082
14	AF417667 *N. lecointei*	0.055	0.094	0.076	0.062	0.115	0.094	0.101	0.071	0.068	0.052	0.073	0.063
15	KX839240 *N. lorenziana*	0.085	0.115	0.097	0.088	0.123	0.108	0.118	0.064	0.052	0.081	0.069	0.090
16	HF679198 *N. soratensis*	0.118	0.099	0.120	0.122	0.065	0.088	0.112	0.104	0.100	0.103	0.115	0.106
17	AF417670 *N. alba*	0.083	0.110	0.087	0.086	0.129	0.109	0.114	0.069	0.077	0.083	0.080	0.082
18	HF679149 *N. capitellata*	0.076	0.107	0.074	0.080	0.117	0.099	0.099	0.026	0.070	0.070	0.056	0.074
19	KX839236 *N. pusilla*	0.080	0.121	0.079	0.086	0.111	0.111	0.097	0.044	0.067	0.076	0.049	0.074
20	KX839243 *N. navisvaringica*	0.019	0.094	0.092	0.027	0.111	0.094	0.110	0.088	0.083	0.074	0.089	0.082
21	KX839237 *Nitzschia* sp1	0.059	0.110	0.089	0.061	0.117	0.109	0.112	0.064	0.063	0.064	0.079	0.086

Table 3 contd. right side

	Species	13	14	15	16	17	18	19	20	21
13	KX839238 *N. amabilis*	-								
14	AF417667 *N. lecointei*	0.064	-							
15	KX839240 *N. lorenziana*	0.078	0.076	-						
16	HF679198 *N. soratensis*	0.106	0.107	0.119	-					
17	AF417670 *N. alba*	0.086	0.074	0.089	0.118	-				
18	HF679149 *N. capitellata*	0.076	0.061	0.066	0.104	0.071	-			
19	KX839236 *N. pusilla*	0.080	0.071	0.069	0.103	0.069	0.044	-		
20	KX839243 *N. navisvaringica*	0.037	0.053	0.087	0.118	0.086	0.077	0.086	-	
21	KX839237 *Nitzschia* sp1	0.056	0.046	0.070	0.116	0.074	0.068	0.071	0.062	-

In the other hand, *N. johorensis* sp. nov. formed an isolated branch giving 9%–13% genetic divergence from other *Nitzschia* species (Table 3). The most morphologically similar species *N. Fonticola* also clustered into another clade and is 12.3% divergence from *N. johorensis* sp. nov. Apart from that, the dimorphic resemblance of *N. alba* did not show clear relationship with *N. johorensis* sp. nov. and there were no genetic data for *N. dubia* and *N. frigida*.

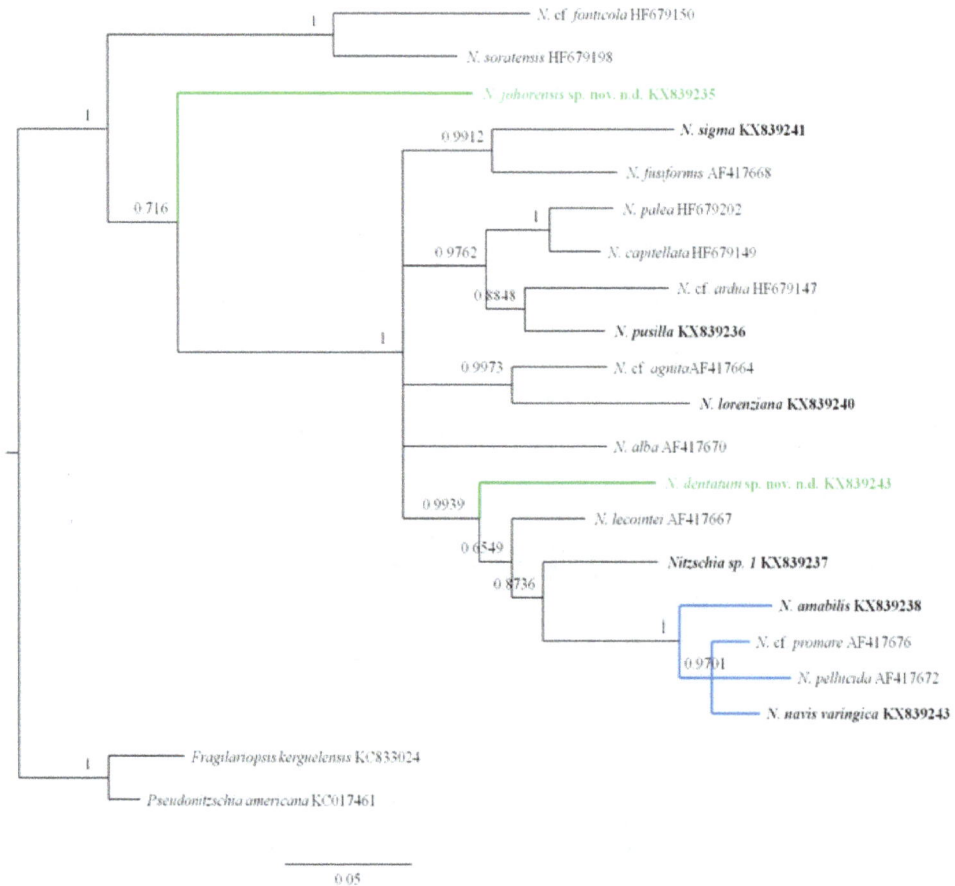

Fig. 4. Phylogenetic trees of *Nitzschia* spp. reconstructed based on the D1–D3 gene region of LSU rDNA using Bayesian Analysis with *P. americana* and *F. kergeulensis* as out groups. Bold-lettered strains were obtained from this study (black) and proposed species (green). Species that are constricted in the middle valve grouped into the same clade consistently in all tress (blue). Prior probability values are shown.

From our observation, entities that have similar outlines tend to group together in the tree. For instance, cells with middle indentations of valves i.e. *N. navis-varingica*, *N. amabilis*, *N. promare*, *N. pellucida* and *N. laevis* clustered together as a group. To our knowledge, there are no other cells that have middle indentations located outside of that clade in the tree. *N. amabilis* is the new nomination of *N. laevis* (Suzuki *et al.*, 2010) and hence is the same species. *N. amabilis* branches out separately in the phylogenetic tree but it differs from *N. laevis* (syn. *N. amabilis*) genetically by 4%. There was no strong evidence to prove them as separate species. *N. polaris* Grunow ex Cleve and *N. neglecta* Hustedt (Medlin and Hasle, 1990) are among others that have indentation in the middle, but there were no DNA sequences available.

The behavior of colony formation also reflects the clustering but was very weakly supported (Lundholm *et al.,* 2002). The strains' habitat, localities, and toxicity were not resolved in the phylogenetic tree. Freshwater species *N. palea* (Kützing) W. Smith and *N. sigma* (Kützing) W. Smith (Aishah and Nooraida, 1994) grouped with other marine species. Likewise, *N. amabilis* was isolated from tropical water (Suriyanti, 2017) while *N. promare* is a polar species (Medlin and Hasle, 1990). In terms of toxicity, the DA producers *N. bizertensis* Smida, Lundholm, Sakka and Hadj Mabrouk and *N. navis-varingica* LundholmetMoestrupwere also not closely related (Smida *et al.,* 2014). Both of the newly proposed species in this study are non-toxic.

It is indeed very difficult to identify the key traits for the phylogenetic grouping of the genus *Nitzschia*. Lengths and widths are not stable characters to differentiate among *Nitzschia* species (Suriyanti, 2017) due to measurements that mostly overlap among species. Other ultra-structural features such as presence of central nodules, densities of striae and fibulae as well as rows of poroids did not show any distinguishable pattern either (Lundholm *et al.,* 2002). The valve outlines, raphe arrangements and presence of central interspace seemed to be persistent in species delineation in this genus (Mann, 1978; Round *et al.,* 1990; Lundholm and Moestrup, 2000). Inconsistent taxa placement in the tree is observed in species such as *N. pusilla*, even though most of the strains from Genbank have been verified by experts (Lundholm *et al.,* 2002). This could indicate the existence of cryptic and pseudo-cryptic species. Further studies are required to develop a better insight of relations between the genetic encoding and the morphology of *Nitzschia*.

From the morphology and genetic data, it is therefore to affirm that *N. dentatum* sp. nov. and *N. johorensis* sp. nov. have not been described elsewhere. The phylogenetic trees inferred usingLSU rDNA gene sequences also revealed *Nitzschia* species that have indentation in the middle of valve were grouped into the same clade consistently. Further genetic analyses are necessary to clarify the natural groupings within the genus *Nitzschia*.

Acknowledgements

Suriyanti S.N.P. was funded under MyBrain15 scholarship by the Malaysian Ministry of Higher Education. We would like to thank Dr. Dzulhelmi Nasir for his guidance in bioinformatics and Mr. Zaki for assisting the SEM operation.

References

Aishah, S. 2005. Phytoplankton. *In*: Sasekumar, A. and Chong, V.C. (Eds),Ecology of Klang Strait. University of Malaya Press, Kuala Lumpur, Malaysia, pp. 133–167.

Aishah, S. and Nooraida, A.R. 1994. Diatoms of the River Kanching, Templer Park, Malaysia. Diatom**9**: 73–81.

Chu, W.L., Phang, S.M. andGoh, S.H. 1996. Environmental effects on growth and biochemical composition of *Nitzschia inconspicua* Grunow. J. Appl. Phycol. **8**: 389–396.

Cleve, P.T. 1901. Plankton from the Indian Ocean and the Malay Archipelago. K. Svenska Vetenskapsakademiens Handl. **35**(5): 1–58.

Coste, M. and Ricard, M. 1980. On some interesting finely striated *Nitzschia* observed under light and electron microscopes. Systematic and ecological Aspects. *In*: Ross, R. (Ed.),Proceedings of the 6th Symposium of Recent Fossil Diatoms.Otto Koeltz Science Publishers, Koenigstein, pp. 191–202.

Fareha, H., Leaw, C.P. and Lim, P.T. 2011. Morphological observation of common pennate diatoms (Bacillariophyceae) from Sarawak estuarine waters. Ann. Microsc. **11**: 12–23.

Foged, N. 1971. Freshwater Diatoms in Thailand. Cramer, Lehre, pp. 1–102.

Guiry, M.D. and Guiry, G.M. 2017. Algae Base. <http://www.algaebase.org>. National University of Ireland, Galway. Retrieved on 29 August 2017.

Hall, T.A. 1999. Bioedit: A user-friendly biological sequence alignment editor and analysis program for Windows 95/98/Nt. Nucl. Acids. Symp. Ser.**41**: 95–98.

Iwasaki, H. 1961. The life-cycle of *Porphyra tenera in vitro*. Biol. Bull.**121**: 173–187.

Ki,, J.S. andHan,M.S. 2005. Molecular analysis of complete SSU to LSU rDNA sequence in the harmful dinoflagellate *Alexandrium tamarense* (Korean Isolate, Hy970328m). Ocean Sci. J. **40**(3): 43–54.

Kociolek, P. 2011. *Nitzschia fonticola*. Diatoms of the United States. <http://westerndiatoms. colorado. edu/ taxa/species/nitzschia_fonticola>. Retrieved on14 December 2015.

Laurittis, J.A., Hemmingsen, B.B. and Volcani, B.E. 1967. Propagation of *Hantzschia* sp. Grunow daughter cells by *Nitzschia alba* Lewin & Lewin. Phycol. **3**: 236–237.

Lee, J.J. 2011. Diatoms as endosymbionts. *In*: Seckbach, J. and Kociolek, J.P. (Eds), The diatom world. Springer Science & Business Media, New York, pp. 439–463.

Lundholm, N., Daugbjerg, N. and Moestrup, Ø. 2002. Phylogeny of the Bacillariaceae with emphasis on the genus *Pseudo-nitzschia* (Bacillariophyceae) based on partial LSU rDNA. Eur. J. Phycol. **37**(1): 115–134.

Lundholm, N. and Moestrup, Ø. 2000. Morphology of the marine diatom *Nitzschia navis-varingica* sp. nov. (Bacillariophyceae), Another producer of the neurotoxin domoic acid. J.Phycol. **36**: 1162–1174.

Mann, D.G. 1978. Studies in the Family Nitzschiaceae (Bacillariophyta) Vols**1 & 2**. Xxxiii + 386 pp. + 146 Plates. Ph.D. Dissertation, Universiti of Bristol, United Kingdom.

Maznah,W.W.O.and Mansor, M. 2002. Aquatic pollution assessment based on attached diatom communities in the Penang River Basin, Malaysia. Hydrobiologia**487**: 229–241.

Medlin, L.K. andHasle, G.R. 1990. Some *Nitzschia* and related diatom species from fast ice samples in the Arctic and Antarctic. Polar Biol.**10**: 451–479.

Nather-Khan, I.S.A. 1990. Assessment of water pollution using diatom community structure and species distribution -A case study in a Tropical River Basin. Int. Rev.Hydrobiol.**75**(3): 317–338.

Nunn, G.B., Theisen, B.F., Christensen, B. and Arctander, P. 1996. Simplicity-correlated size growth of the nuclear 28s ribosomal RNA D3 expansion segment in the Crustacean Order Isopoda. J. Mol.Evol. **42**: 211–223.

Posada, D. and Crandall, K.A. 1998. Modeltest: testing the model of DNA substitution. Bioinformatics **14**(9): 817–818.

Renberg, I. 1990. A procedure for preparing large sets of diatom slides from sediment cores. J. Paleolimnol. **4**: 87–90.

Ronquist, F. and Huelsenbeck, J.P. 2003. MrBayes3: Bayesian phylogenetic inference under mixed models. Bioinformatics **19**: 1572–1574.

Round, F.E, Crawford, R.M. and Mann, D.G. 1990. The Diatoms: Biology and Morphology of the Genera. Cambridge University Press, New York, USA, 747pp.

Saifullah, A.S.M, Abu Hena, M.K., Idris, M.H., Halimah, A.R. and Johan, I. 2014. Diversity of phytoplankton from mangrove estuaries of Sarawak, Malaysia. World Appl. Sci. J. **31**(5): 915–924.

Scholin, C.A., Herzog, M., Sogin, M. and Anderson, D.M.1994. Identification of group- and strain-Specific genetic markers for globally distributed *Alexandrium* (Dinophyceae). II. Sequence analysis of a fragment of the LSU rDNA. J. Phycol. **30**: 999–1011.

Shamsudin, L. 1990. Diatom marin di perairan Malaysia. Percetakan Dewan Bahasadan Pustaka, Kuala Lumpur, Malaysia, 260 pp.

Smida, D.B., Lundholm, N., Kooistra, W.H.C.F., Sahraoui, I., Ruggiero, M.V., Kotaki, Y., Ellegaard, M., Lambert, C., Mabrouk, H.H. and Hlaili, A.S. 2014. Morphology and molecular phylogeny of *Nitzschia bizertensis* sp. nov. - A new domoic acid-producer. Harmful Algae **32**: 49–63.

Sonnenberg, R., Nolte, A.W. and Tautz, D. 2007. An evaluation of LSU rDNA D1–D2 sequences for their use in species identification. Front. Zool. **4**: 6.

Suriyanti, S.N.P. 2017. Kajiantaksonomi, filogenidanketoksikandiatom marin *Nitzschia* (Bacillariophyceae) di perairan Malaysia. PhD Thesis. Universiti Kebangsaan Malaysia, Bangi, Malaysia.

Suriyanti, S.N.P. and Gires, U. 2015. First report of the toxigenic *Nitzschianavis-varingica* (Bacillariophyceae) isolated from Tebrau Straits, Johor, Malaysia. Toxicon**108**: 257–263.

Suzuki, H., Nagumo, T. and Tanaka, J. 2010. *Nitzschiaamabilis* nom. nov., a new name for the marine species *N. Laevis* Hustedt. Diatom Res. **25**(1): 223–224.

Swofford, D.L. 2003. PAUP*: Phylogenetic analysis using parsimony (*and other methods). Version 4.0b10. Sinauer Associates, Sunderland.

Tamura, K., Stecher, G., Peterson, D., Filipski, A. and Kumar, S. 2013. MEGA6: Molecular evolutionary genetics analysis version 6.0. Mol. Biol. Evol. **30**: 2725–2729.

Thompson, J.D., Gibson, T.J., Plewniak, F., Jeanmougin, F. and Higgins, D.G. 1997. The clustalX windows interface: Flexible strategies for multiple sequence alignment aided by the quality analysis tools. Nucleic Acid Res. **24**: 4876–4882.

Trobajo, R., Cox, E.J. and Quintana, X.D. 2004. The Effects of some environmental variables on the morphology of *Nitzschia frustulum* (Bacillariophyta), in relation to its use as a bioindicator. Nova Hedwigia**79**: 433–445.

Trobajo, R., Clavero, E., Chepurnov, V.A., Sabbe, K., Mann, D.G., Ishihara, S. and Cox, E.J. 2009. Morphological, genetic and mating diversity within the widespread bioindicator *Nitzschiapalea* (Bacillariophyceae). Phycologia **48**(6): 443–459.

Trobajo, R., Rovira, L., Ector, L., Wetzel, C.E., Kelly, M. and Mann, D.G. 2013. Morphology and identity of some ecologically important small *Nitzschia* species. Diatom Res. **28**(1): 37–59.

Tudesque, L., Rimet, F. andEctor, L. 2008. A new taxon of the Section *Nitzschiae Lanceolatae* Grunow: *Nitzschia costei* sp. nov. compared to *N. fonticola* Grunow, *N. macedonia* Hustedt, *N. tropica* Hustedt and related species. Diatom Res. **23**(2): 483–501.

Zhang, Z., Schwartz, S., Wagner, L. and Miller, W. 2000. A greedy algorithm for aligning DNA sequences. J.Comput. Biol. **7**(1–2): 203–14.

Zimmermann, J., Jahn, R. and Gemeinholzer, B. 2011. Barcoding diatoms: Evaluation of the V4 Subregion on the 18s rRNA Gene, including new primers and protocols. Org. Divers. Evol. DOI 10.1007/s13127-011-0050-6

15

MOLECULAR IDENTIFICATION OF *LAVENDULA DENTATA* L., *MENTHA LONGIFOLIA* (L.) HUDS. AND *MENTHA × PIPERITA* L. BY DNA BARCODES

Shawkat Mahmoud Ahmed[1]

Biology Department, Faculty of Education, Ain Shams University, Cairo, Egypt

Keywords: *Lavendula*; *Mentha*; ITS; ITS2; *rbc*L; *mat*K; *trn*H.

Abstract

Five DNA barcodes were tested for identification and discrimination of *Lavendula dentata* L., *Mentha longifolia* (L.) Huds. and *Mentha × piperita* L. New DNA barcodes have been registered for *L. dentata* from Taif, Saudi Arabia. The separate clading of *L. dentata* and *M. longifolia* through the phylogenic analyses proved their endemism to Saudi Arabia. The phylogenetic trees revealed from the ITS2, *mat*K and *trn*H data demonstrated that all *Mentha* species formed monophyletic clusters except hybrid *M. × piperita* from Taif which formed separate clades distinguishing it from the two parents; *M. aquatica* L. and *M. spicata* L. DNA barcoding could be considered as a good approach for distinguishing and identifying the mint plants, though it was not possible to confirm the relationship between hybrids and their putative parents.

Introduction

The family Lamiaceae comprising about 7,173 species under 236 genera possesses medicinal and aromatic herbs such as lavender, basil, mint, rosemary and thyme, that have been widely utilized as teas, spices, traditional medicines or raw material for the food and pharmaceutical industries (Theodoridis *et al.*, 2012).

Lavendula dentata is one of five naturally growing lavender species in Saudi Arabia that has been known as the main center of origin of the genus (Miller, 1985). Lavender species as medicinal plants, are distributed in highlands of Albaha, Asir, and Taif and are exploited for the production of high-quality lavender honeys. Locally known as Habak, Al-Madinah mint or wild mint, *Mentha longifolia* and peppermint, *Mentha × piperita* are present in the spontaneous flora of Saudi Arabia but also under cultivation. Traditionally, they have been used as medicinal agents to treat colds, cough, headaches, asthma and digestive disorders. Recent studies proved the antiviral, antimicrobial, antioxidant, anti-inflammatory and anticancer characteristics as therapeutic activities for the extracts derived from *Mentha* species (Anwar *et al.*, 2017). Hybridization and polyploidy play an important role in the speciation of the members belonging to genus *Mentha* such as *M. × piperita* that is considered as a hybrid of the two mints; *M. spicata* and *M. aquatica* (Mogosan *et al.*, 2017) making them good targets for molecular studies.

Various studies have been performed to identify and classify species of Lamiaceae collected from Saudi Arabia based on anatomical and cytological studies (Abdel Khalik, 2016) and biochemical analyses (Kasem, 2016), however, very little is known about DNA barcoding information. DNA sequences for the species under study will be compared in a database against retrieved sequences of identified individuals from the GenBank. If the query sequence matches with one in the database, this will help in identification, discrimination or gaining a new barcodes for these species (Hajibabaei *et al.*, 2007). Therefore, the objectives of this research include:

---

[1]Present address: Biology Department, Faculty of Science, Ta'if University, Ta'if, 5700, Saudi Arabia. Email: shamahmoh@gmail.com

i) utility of specific DNA regions, two nuclear internal transcribed spacers (ITS and ITS2) and the plastid DNA regions (*rbc*L, *mat*K and *trn*H) for developing DNA barcodes and subsequently identification for the three species; *Lavendula dentata, Mentha longifolia* and *M.* × *piperita* occurred in Taif highlands of Saudi Arabia; ii) discriminating between species under study and those retrieved from the GenBank and iii) exploring the interspecific variation between *M. longifolia* and *M.* × *piperita*.

Materials and Methods
Plant materials
Two wild species, namely *Lavendula dentata* and *Mentha longifolia,* and the hybrid species, *M.* × *piperita* belonging to family Lamiaceae were collected from Taif highlands, Saudi Arabia. Species identification was confirmed following Collenette (1999).

DNA extraction and amplification
DNA of fresh young leaves was extracted using CTAB method as described by Doyle and Doyle (1987). The purified DNA was amplified for ITS, ITS2, *rbc*L, *mat*K and *trn*H barcodes using universal primers.

PCR sequencing
The PCR products of the three Lamiaceae species for the five DNA barcodes were purified and sequenced at Macrogen Inc., South Korea. All sequences of the three species generated in this research were deposited in GenBank (accession numbers are listed in Table 1).

Sequences alignment and phylogenetic analyses
The sequences of ITS, ITS2, *mat*K, *rbc*L and *trn*H of *L. dentata, M. longifolia* and *M.* × *piperita* were subjected to BLAST (http://blast.ncbi.nlm.nih.gov/Blast.cgi) to confirm them from the other related Lamiaceae species existing in the GenBank database. Sequence alignments were performed by MUSCLE algorithm (Edgar, 2004; Tamura *et al.,* 2013). The equality of evolutionary rate parameters between sequences of the three species under study and the retrieved species from GenBank were calculated by Tajima's relative rate test (Tajima, 1993). Nucleotide substitution rates and Transition/Transversion bias (R) were estimated using Maximum Likelihood method. The phylogenic trees were constructed by the Maximum likelihood bootstrap (MLB) analysis. A total of 1,000 bootstrap replicates were performed. The software of MEGA6 was used for all operations (Tamura *et al.,* 2013).

Table 1. Accession numbers in GenBank of sequences of *Lavendula dentata, Mentha longifolia* and *M.* × *piperita* generated in this study.

Taxa	ITS	ITS2	*mat*K	*rbc*L	*trn*H
L. dentata	LC373552.1	LC373553.1	-	LC373554.1	LC373555.1
M. longifolia	-	LC378378.1	-	LC378379.1	-
M. × *piperita*	-	LC374287.1	LC374288.1	LC374289.1	LC374290.1

Results and Discussion
Identification of Lavendula dentata
Sequences of *L. dentata* for ITS, ITS2, *rbc*L and *trn*H barcoding loci were submitted to BLAST at the GenBank database, however, any sequence of *L. dentata* was detected in the database, thus the present study succeeded in registering new DNA barcodes for *L. dentata* from

Taif. Sequences of species belonging to the genus *Lavendula* showing high similarities to those of *L. dentata* were retrieved for the statistical analyses. ITS showed the highest sequence length (775 bp) followed by *rbc*L (537 bp), ITS2 (358 bp) and *trn*H (346 bp), whereas, the variable sites percentage after alignment was higher in *trn*H (24%) than those of ITS, ITS2 and *rbc*L. The GC ratios scored in loci ITS and ITS2 (60.1 and 65.9) was found greater than those of *rbc*L and *trn*H (Table 2). In comparison with the retrieved *Lavendula* species, the rates of transitions to transversions showed notable substitution changes in the sequences of *L. dentata* (Table 2). Transitions generally occurred more than transversions. Transition/transversion bias (R) was found relatively high and ranged from 1.19 to 2.82 demonstrating a molecular evolution within *Lavendula* genome. This putative evolution in *L. dentata* was confirmed through tests of Tajima relative evolutionary rate that displayed an accelerated rates of evolution (*P*-values <0.05) for all loci under study. The results revealed that ITS, ITS2, *rbc*L and *trn*H have sufficient efficiency in sequence quality as well as in species identification across the genome of the genus *Lavendula*. For further identification of *L. dentata*, sequences of the four loci were used to reconstruct four phylogenetic trees (Fig. 1). Except the tree revealed from ITS2, the separate clustering of *L. dentata* in the phylogenetic trees of ITS, *rbc*L and *trn*H proved its endemism to Saudi Arabia. The development of different DNA barcodes is better than single locus for more accurate results (Khan *et al.*, 2013). The identification of species within a community through DNA barcodes contributes to the construction of the barcode library for terrestrial plants (Burgess *et al.*, 2011).

Table 2. Statistics derived from the sequencing, alignment and BLAST processes for all loci employed in the present investigation.

Parameters	Loci				
	ITS	ITS2	*rbc*L	*mat*K	*trn*H
% Variable sites after alignment for *Lavendula dentata*	0.01	0.06	0.01		0.24
% Variable sites after alignment for *Mentha longifolia*	0.11	0.12	0.04	-	-
% Variable sites after alignment for *M. × piperita*	-	0.12	0.04	0.29	0.31
Sequence length of *L. dentata*	775	358	537	-	346
Sequence length of *M. longifolia*	362	347	528	-	-
Sequence length of *M. × piperita*	-	349	540	810	403
GC ratio in *L. dentata*	60.1	65.9	43.6	-	28.6
GC ratio in *M. longifolia*	51.1	66.6	44.1	-	-
GC ratio in *M. × piperita*	-	67.9	43.7	34.8	31.2
Number of the retrieved *Lavendula* species from the GenBank	3	2	4	-	2
Number of the retrieved *Mentha* species from the GenBank	4	9	9	11	14

Identification of Mentha longifolia

Sequences of ITS, ITS2 and *rbc*L were used to identify *M. longifolia*. ITS2 recorded the lowest sequence length, whereas, the variable sites (%) and GC ratio of it were greater than those of ITS and *rbc*L (Table 2). Transitions were found to be more than transversions leading to substitution changes in the sequences of *M. longifolia* (Table 3). An evolution within *M. longifolia* genome was noticed through the high transition/transversion bias (R) that ranged from 3.51 in *rbc*L to 1.81 in ITS2. Except data of *rbc*L, Tajima relative evolutionary rate displayed an accelerated rates of evolution (*P*-values <0.05) in *M. longifolia* (Table 4). Sequences of *M. longifolia* for ITS, ITS2 and *rbc*L that submitted to BLAST at the GenBank retrieved 4, 9 and 9 *Mentha* species, respectively (Table 2). *M. longifolia* and the retrieved *Mentha* species reconstructed three phylogenetic trees (Fig. 2) which revealed that *M. longifolia* was represented in separate clade demonstrating variability between it and other *Mentha* species, and proved its endemism to Saudi Arabia. Similar result was obtained by Khan *et al.* (2013) in *Senecio asirensis* using nrDNA ITS.

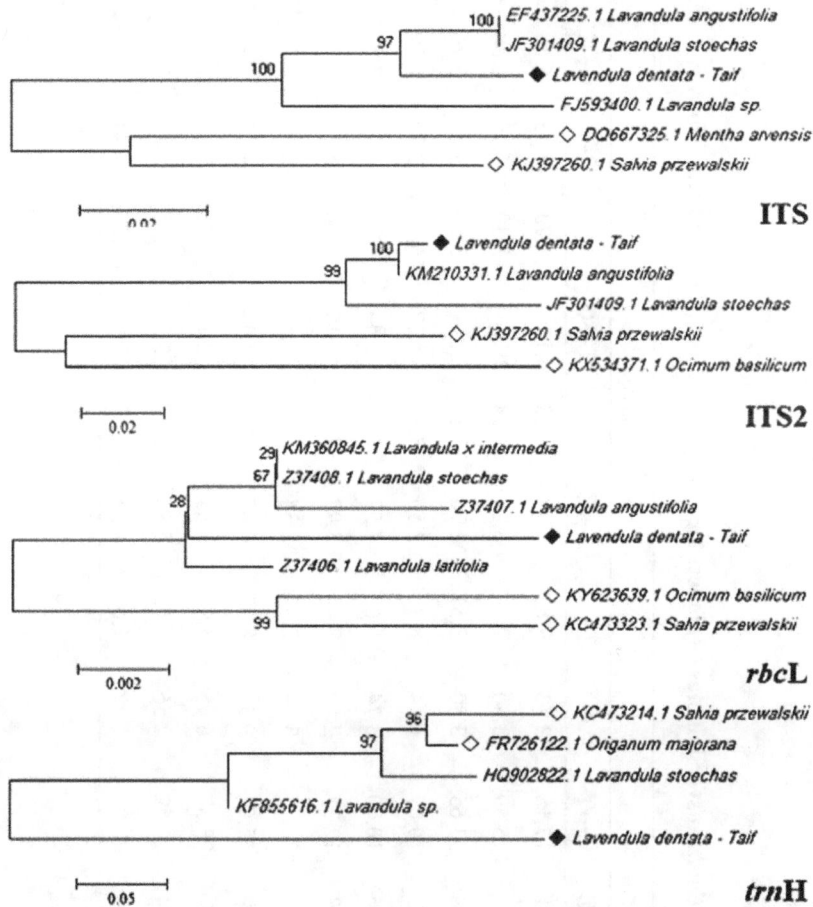

Fig. 1. Phylogenetic trees of *Lavendula dentata* and the retrieved species based on four loci. (◊) refers to the outgroup.

Identification of M. × piperita

Sequences of ITS2, *rbc*L, *mat*K and *trn*H were used to identify *M. × piperita*. As found in *M. longifolia*, ITS2 showed the lowest sequence length (349 bp) and the highest GC ratio (67.9). Whereas, the percentage of variable sites (31%) was detected in *trn*H locus (Table 2). An obvious evolution was also observed within *M. × piperita* genome through the high transition/transversion bias (R) that ranged from 0.79 in *mat*K to 7.01 in *rbc*L (Table 3). The previous result was supported by Tajima relative evolutionary rate that displayed an accelerated rates of evolution (*P*-values <0.05) in *M. × piperita* except that of *rbc*L (Table 4). The null hypothesis of equal evolution rates between *M. × piperita* from Taif and its ancestors; *M. spicata* and *M. aquatica* from one hand, and the retrieved *M. × piperita* from the other hand, was rejected because the *P*-values were lower than 0.05 in ITS2, *mat*K and *trn*H revealing the accelerated evolutionary rate of *M. × piperita* from Taif and subsequently reflecting the variance among them. *M. × piperita* and the retrieved *Mentha* species from the GenBank library were analyzed to form four phylogenetic trees (Fig. 2). The phylogenetic trees from the ITS2, *mat*K and *trn*H data demonstrated that all the

Table 3. Mean nucleotide substitution rates and transition/transversion bias (R) in loci for *L. dentata*, *M. longifolia* and *M.* × *piperita*.

Taxa	Locus	Transition				Transversion								(R)
		A→G	G→A	T→C	C→T	A→T	T→A	A→C	C→A	T→G	G→T	C→G	G→C	
L. dentata	ITS	7.75	4.59	30.31	17.18	3.73	3.63	6.58	3.63	6.13	3.73	6.13	6.58	1.40
	ITS2	14.68	5.81	29.66	15.59	3.30	2.14	6.28	2.14	29.66	3.30	5.40	6.28	1.79
	*rbc*L	7.19	8.78	24.48	33.65	3.80	3.51	2.76	3.51	2.87	3.80	2.87	2.76	2.82
	*trn*H	9.07	23.09	7.57	19.07	6.44	8.32	2.56	8.32	3.27	6.44	3.27	2.56	1.19
M. longifolia	ITS	11.20	11.64	21.66	10.70	3.74	5.65	7.57	5.65	5.44	3.74	5.44	7.57	1.15
	ITS2	17.14	8.07	30.52	12.37	2.40	2.44	5.92	2.44	5.19	2.40	5.19	5.92	1.81
	*rbc*L	15.21	17.97	2.51	25.75	3.14	2.97	2.34	2.97	2.51	3.14	2.51	2.34	3.51
M. x piperita	*mat*K	8.27	15.50	7.48	15.62	9.34	8.32	4.47	8.32	4.44	9.34	4.44	4.47	0.79
	*trn*H	8.44	23.85	7.16	28.37	6.45	5.92	1.63	5.92	2.10	6.45	2.10	1.63	1.48
	ITS2	17.25	8.10	28.84	11.59	2.56	2.62	6.36	2.62	5.57	2.56	5.57	6.36	1.63
	*rbc*L	17.09	20.19	21.49	28.93	1.77	1.67	1.31	1.67	1.41	1.77	1.41	1.31	7.01

Mentha species formed monophyletic clusters except the hybrid *M. × piperita* from Taif which formed separate clades. The differences between *M. × piperita* under study and the other retrieved *Mentha* species could be explained due to an evolutionary process.

Little divergence in *rbc*L tree (Fig. 2) and the acceptance of the null hypothesis of equal evolutionary rates among *Mentha* species through *rbc*L data (Table 4) could be due to the symmetry in *rbc*L sequence of *Mentha* species. Kshirsagar *et al.* (2015) reported the same limitation of *rbc*L gene in closely related species of the two genera *Ardisia* Sw. and *Swertia* L. These results were in accordance with those of Theodoridis *et al.* (2012) who showed that *mat*K and *trn*H were more useful in discriminating Lamiaceae species than *rbc*L. It was noticed that ITS2, *rbc*L, *mat*K and *trn*H distinguished *M. × piperita* from the two parents, *M. aquatica* and *M. spicata* through the phylogentic trees. These genetic differences might be due to most commercial

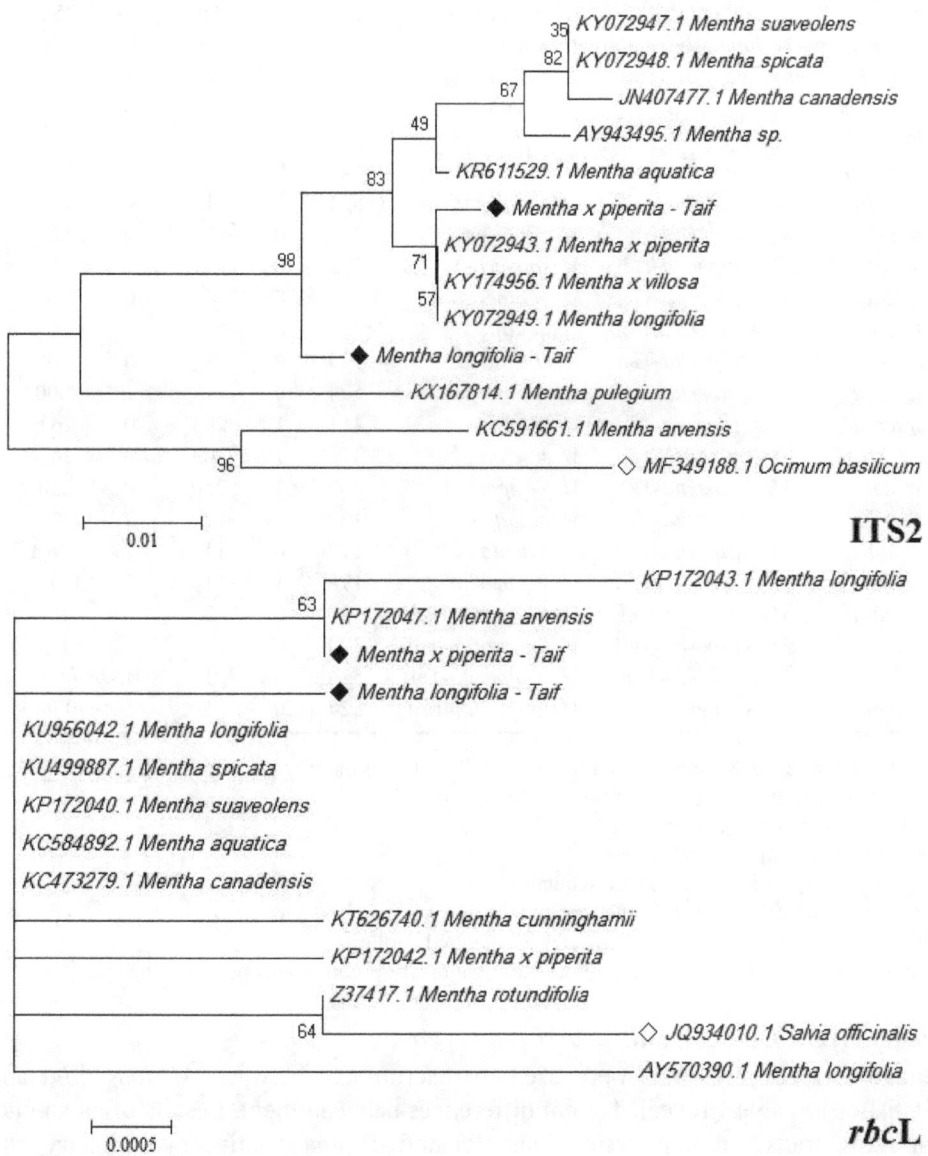

Fig. 2. Phylogenetic tree diverges between *Mentha longifolia* and *M. × piperita* based on ITS2 and *rbc*L sequences. (◊) refers to the outgroup.

hybrids, i.e. *M. × piperita* is sterile or subfertile, therefore, crossing with parental or nonparental species is expected. They may also form complex hybrid populations through vegetative propagation and polyploidy. These possibilities lead to great genetic diversity and subsequently to several taxonomic problems (De Mattia *et al.*, 2011).

Table 4. Tajima relative rate tests of loci for *L. dentata*, *M. longifolia* and *M. × piperita*.

Loci	Outgroup	Testing group		RI	RD	RA	RB	χ2	P value
		(A)	(B)						
ITS	*L. angustifolia*	*L. dentata*-Taif	*L. stoechas*	512	0	36	0	36.0	<0.05
ITS2	*L. angustifolia*	*L. dentata*-Taif	*L. stoechas*	280	0	2	17	11.8	<0.05
*rbc*L	*L. angustifolia*	*L. dentata*-Taif	*L. stoechas*	529	0	5	0	5.00	<0.05
*trn*H	*L. angustifolia*	*L. dentata*-Taif	*L. stoechas*	126	6	60	16	25.5	<0.05
ITS	*M. suaveolens*	*M. longifolia*-Taif	*M. spicata*	70	0	33	0	33.0	<0.05
	M. spicata	*M. longifolia*-Taif	*M. suaveolens*	70	0	33	1	30.12	<0.05
ITS2	*M. × piperita*	*M. longifolia*-Taif	*M. longifolia*	297	0	9	0	9.00	<0.05
	M. longifolia	*M. longifolia*-Taif	*M. × piperita*	297	0	9	0	9.00	<0.05
*rbc*L	*M. × piperita*	*M. longifolia*-Taif	*M. longifolia*	524	0	1	0	1.00	>0.05
	M. longifolia	*M. longifolia*-Taif	*M. × piperita*	524	0	1	1	0.00	>0.05
ITS2	*M. spicata*	*M. × piperita*-Taif	*M. aquatica*	272	0	7	0	7.00	<0.05
	M. aquatica	*M. × piperita*-Taif	*M. spicata*	272	0	7	0	7.00	<0.05
	M. spicata	*M. × piperita*-Taif	*M. × piperita*	296	0	4	0	4.00	<0.05
*rbc*L	*M. spicata*	*M. × piperita*-Taif	*M. aquatica*	525	0	1	0	1.00	>0.05
	M. aquatica	*M. × piperita*-Taif	*M. spicata*	525	0	1	0	1.00	>0.05
	M. spicata	*M. × piperita*-Taif	*M. × piperita*	524	0	1	1	0.00	>0.05
*mat*K	*M. spicata*	*M. × piperita*-Taif	*M. aquatica*	411	1	263	0	263	<0.05
	M. aquatica	*M. × piperita*-Taif	*M. spicata*	411	1	263	0	263	<0.05
	M. spicata	*M. × piperita*-Taif	*M. × piperita*	400	1	250	0	250	<0.05
*trn*H	*M. spicata*	*M. × piperita*-Taif	*M. aquatica*	177	0	114	2	108.1	<0.05
	M. aquatica	*M. × piperita*-Taif	*M. spicata*	177	0	114	2	108.1	<0.05
	M. spicata	*M. × piperita*-Taif	*M. × piperita*	178	1	114	1	111.0	<0.05
ITS2	*M. spicata*	*M. longifolia*-Taif	*M. × piperita*-Taif	320	0	3	6	1.0	>0.05
	M. spicata	*M. × piperita*-Taif	*M. longifolia*-Taif	320	0	3	6	1.0	>0.05
*rbc*L	*M. spicata*	*M. longifolia*-Taif	*M. × piperita*-Taif	524	0	1	1	0.00	>0.05
	M. spicata	*M. × piperita*-Taif	*M. longifolia*-Taif	524	0	1	1	0.00	>0.05

The Tajima relative rate test was used to examine the equality of evolutionary rate for *L. dentata*, *M. longifolia* and *M. × piperita* and other relative species with different outgorups.
RI is the identical sites in all three sequences
RD is the divergent sites in all three sequences
RA is the number of unique differences in the sequence A
RB is the number of unique differences in the sequence B
χ2 test statistic more than 3.841 (*P* <0.05) indicates accelerated evolution
P value greater than 0.05 is often used to accept the null hypothesis of equal rates between lineages

Discrimination between M. longifolia and M × piperita

　　　Sequences of ITS2 and *rbc*L were used to discriminate between *M. longifolia* and *M × piperita*. Statistics in Table 1 revealed slight differences between them. Results of mean nucleotide substitution rates, transition/transversion bias (R) and Tajima relative evolutionary rate were similar in these two taxa (Tables 3 & 4). ITS2 and *rbc*L trees were used to assess genetic divergences between *M. longifolia* and *M × piperita*. A suitable divergence was detected in the two phylogentic trees displaying the efficacy of the two barcodes in distinguishing between them through the Maximum Likelihood method (Fig. 2). Thakur *et al.* (2016) stated that the convenient barcode exhibits large interspecific but little intraspecific divergence and its sequence length must be short enough to be available in a single amplification. This comparability of interspecific sequence variation is a significant aspect for barcoding identification of species in local floras. Establishing a local barcode data will be useful in several ecological applications, such as the reconstruction of community phylogenies, palaeoecological studies of ecosystems and analyzing the diets of human and other animals (Valentini *et al.,* 2009). Depending on these data, the DNA barcoding could be considered as a good approach for distinguishing and identifying the mint plants, however, it was not possible to confirm the relationship between hybrids and their putative parents.

Finally, it could be concluded that the identification and discrimination of *L. dentata, M. longifolia* and *M × piperita* were necessary and valuable for their great economic importance. ITS, *mat*K and *trn*H were found to be more effective barcodes than ITS2 and *rbc*L for the authentication of these species and hybrid. DNA barcoding provided new insight that will contribute to the taxonomy of Lamiaceae taxa around the world and the conservation of the genetic resources of these valuable taxa occurring in Saudi Arabia.

Acknowledgement

The technical support from Macrogen Inc., South Korea is gratefully acknowledged.

References

Abdel Khalik, K.N. 2016. A Systematic revision of the genus *Plectranthus* L. (Lamiaceae) in Saudi Arabia based on morphological, palynological and micromorphological characters of trichomes. Am. J. Plant Sci. **7**: 1429–1444.

Anwar, F., Alkharfy, K.M., Rehman, N., Adam, E.H.K. and Gilani, A. 2017. Chemo-geographical variations in the composition of volatiles and the biological attributes of *Mentha longifolia* L. essential oils from Saudi Arabia. Int. J. Pharmacol. **13**: 408–424.

Burgess, K.S., Fazekas, A.J., Kesanakurti, P.R., Graham, S.W., Husband, B.C., Newmaster, S.G., Percy, D.M., Hajibabaei, M. and Barrett, S.C.H. 2011. Discriminating plant species in a local temperate flora using the *rbc*L+*mat*K DNA barcode. Methods Ecol. Evol. **2**: 333–340.

Collenette, S. 1999. Wild Flowers of Saudi Arabia. National Commission for Wildlife Conservation and Development (NCWCD), Riyadh.

De Mattia, F., Bruni, I., Galimberti, A., Cattaneo, F., Casiraghi, M. and Labra, M. 2011. A comparative study of different DNA barcoding markers for the identification of some members of Lamiaceae. Food Res. Int. **44**: 693–702.

Doyle, J.J. and Doyle, J.L. 1987. A rapid DNA isolation procedure for small quantities of fresh leaf tissue. Phytochem. Bull. **19**: 11–15.

Edgar, R.C. 2004. MUSCLE: multiple sequence alignment with high accuracy and high throughput. Nucleic Acids Res. **32**(5): 1792–1997.

Hajibabaei, M., Singer, G.A.C., Hebert, P.D.N. and Hickey, D.A. 2007. DNA barcoding: how it complements taxonomy, molecular phylogenetics and population genetics. Trends Genet. **23**: 167–172.

Kasem, W.T. 2016. Pollen grains and seed morphology as related to biochemical patterns in five species of genus *Ocimum* L. (Lamiaceae Juss.) of Saudi Arabia. J. Phytol. **8**: 42–48.

Khan, S., Al-Qurainy, F., Nadeem, M. and Tarroum, M. 2013. Selection of chloroplast DNA markers for the development of DNA barcode and reconstruction of phylogeny of *Senecio asirensis* Boulos and J.R.I. wood. Pak. J. Bot. **45**(2): 703–710.

Kshirsagar, P., Umdale, S., Chavan, J. and Gaikwad, N. 2015. Molecular authentication of medicinal plant, *Swertia chirayita* and its adulterant species. Proc. Natl. Acad. Sci. India Sect. B, Biol. **87**(1): 1–7.

Miller, A.G. 1985. The genus *Lavendula* in Arabia and Tropical NE Africa. Notes Roy. Bot. Gard. Edinburgh **42**(3): 503–528.

Mogosan, C., Vostinaru, O., Oprean, R., Heghes, C., Filip, L., Balica, G. and Moldovan, R. 2017. A comparative analysis of the chemical composition, anti-inflammatory, and antinociceptive effects of the essential oils from three species of *Mentha* cultivated in Romania. Molecules **22**: 263–274.

Tajima, F. 1993. Simple methods for testing molecular clock hypothesis. Genetics **135**: 599–607.

Tamura, K., Stecher, G., Peterson, D., Filipski, A. and Kumar, S. 2013. MEGA6: Molecular evolutionary genetics analysis version 6.0. Mol. Biol. Evol. **30**: 2725–2729.

Thakur, V., Tiwari, S., Tripathi, N., Tiwari, G. and Sapre, S. 2016. DNA barcoding and phylogenetic analyses of *Mentha* species using *rbc*L sequences. Ann. Phytomed. **5**(1): 59–62.

Theodoridis, S., Stefanaki, A., Tezcan, M., Aki, C., Kokkini, S. and Vlachonasios, K.E. 2012. DNA barcoding in native plants of the Labiatae (Lamiaceae) family from Chios Island (Greece) and the adjacent Cesme-Karaburun Peninsula (Turkey). Mol. Ecol. Resour. **12**: 620–633.

Valentini, A., Miquel, C., Nawaz, M., Bellemain, E., Coissac, E., Pompanon, F., Gielly, L., Cruaud, C., Nascetti, G., Wincker, P., Swenson, J. and Taberlet, P. 2009. New perspectives in diet analyses based on DNA barcoding and parallel pyrosequencing: the *trn*L approach. Mol. Ecol. Resour. **9**: 51–60.

16

GENETIC VARIATION AMONG IRANIAN ALFALFA (*MEDICAGO SATIVA* L.) POPULATIONS BASED ON RAPD MARKERS

Fatemeh Mohammadzadeh[*], Hassan Monirifar[1], Jalal Saba, Mostafa Valizadeh[2], Ahmad Razban Haghighi[1], Bahram Maleki Zanjani, Maryam Barghi[1] and Vahideh Tarhriz[3]

Faculty of Agriculture, Zanjan University, Zanjan, Iran

Keywords: Alfalfa; RAPD; Genetic diversity; Analysis of Molecular Variance; Cluster analysis.

Abstract

Genetic diversity among and within 10 populations of Iranian alfalfa, from different areas of Azarbaijan, Iran was analyzed by screening DNA from seeds of individual plants and bulk samples. In individual study, 10 randomly amplified polymorphic DNA (RAPD) primers produced 156 polymorphic bands and a high level of genetic diversity was observed within populations. The averages of total and within population genetic diversity were 0.2349 and 0.1892, respectively. Results of analysis of molecular variance (AMOVA) showed the great genetic variation existed within populations (81.37%). These Results were in agreement with allogamous and polyploid nature of alfalfa. Cluster analysis was performed based on Nei's genetic distances resulting in grouping into 3 clusters which could separate breeding population from other populations. Results of cluster analysis were in consistent with morphological and geographical patterns of populations. The results of bulk method were different from individual analysis. Our results showed that RAPD analysis is a suitable method to study genetic diversity and relationships among alfalfa populations.

Introduction

Alfalfa (*Medicago sativa* L.) is the most important forage legume (Veronesi *et al.*, 2010), originated in Caucasus, northeastern Turkey, northwestern Iran and Turkmenistan (Dehghan-Shoar *et al.*, 1997), though Iran is known as central origin (Hanson, 1988). It is an autotetraploid and allogamous plant (Flajoulot *et al.*, 2005). These features lead to its high genetic complexity (Gherardi *et al.*, 1998; Flajoulot *et al.*, 2005). Therefore, a high degree of genetic diversity can be found within and between populations (Mengoni *et al.*, 2000). These factors cause the complication of breeding improvement in alfalfa (Gherardi *et al.*, 1998). However, since alfalfa is an agronomically important crop, its improvement is necessary, especially to increase pest or disease resistance, forage quality and forage yield (Volence *et al.*, 2002). Alfalfa cultivars are synthetic varieties developed by intercrossing the selected parents and advancing their offspring through three or four generations of seed increase (Rowe and Hill, 1999). So, genetic studies such as differentiation between cultivars and estimating the genetic diversity within and between populations are important in alfalfa breeding programs to use some of these populations as selected parents and producing higher yielding cultivars (Veronesi *et al.*, 2010).

---

*Corresponding author: E-mail: fidafeh_m@yahoo.com
[1]Agricultural Biotechnology Research, Institute of Iran (ABRII) for Northwest and West of Iran, Tabriz, Iran.
[2]Faculty of Agriculture, Tabriz University, Tabriz, Iran.
[3]Sari Agricultural Sciences and Natural Resources University, Sari, Iran.

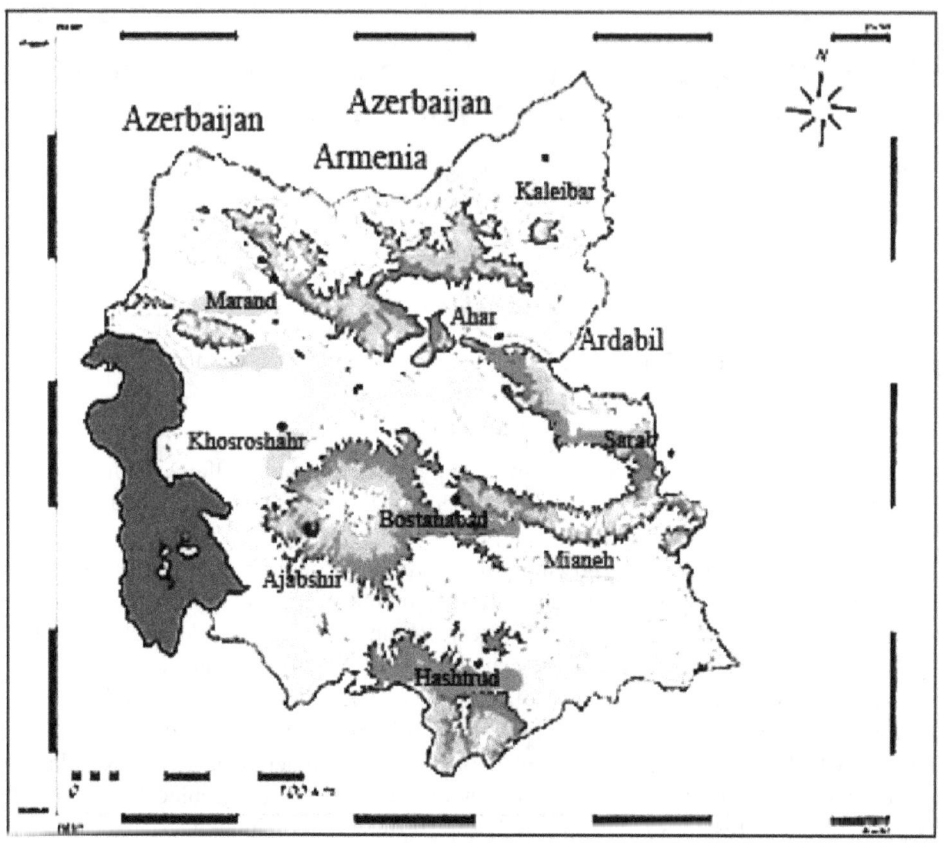

Fig. 1. Locations of alfalfa populations (Azarbaijan, Iran). Source: http://www.ncc.org.ir

DNA-based molecular markers such as RFLPs, SSRs and RAPDs are extensively used to estimate genetic diversity and establish the relationships between plant cultivars (Kidwell *et al.*, 1994; Mengoni *et al.*, 2000). These markers have more polymorphism loci than other methods such as isozyme analysis (Jenczewski *et al.*, 1999) and since they are not affected by environment conditions and plant development level, they can estimate genetic diversity in populations more precisely (Tucak *et al.*, 2008).

In RAPD-PCR technique, genomic DNA is amplified with arbitrary 10-mer oligonucleotide primers to produce DNA fragment polymorphisms (Gherardi *et al.*, 1998). RAPD markers are independent of DNA quantity (Jenczewski *et al.*, 1999) and they do not require previous knowledge of genome (Rahman, 2006; Tucak *et al.*, 2008). Therefore, RAPD analysis is considered as rapid, simple and inexpensive method (Williams *et al.*, 1990; Rahman, 2006) to study genetic structures such as genome mapping, estimating of genetic diversity within and among populations and discriminating among plant populations and cultivars such as alfalfa (Arzani and Samei, 2004; Vandemark *et al.*, 2006; Rahman, 2010).

Although RAPD procedure is a useful method, its application might be limited when a large number of individuals are studied (Yu and Pauls, 1993). This problem could be solved using bulked DNA samples as DNA templates in RAPD amplifications (Michelmore *et al.*, 1991). In this study we aimed to estimate genetic diversity within and among alfalfa populations of Azarbaijan (Iran) by RAPD markers. We also grouped these populations with analysis of individual samples and bulked DNA samples.

Materials and Methods

Plant materials: Nine tetraploid Iranian alfalfa native ecotypes collected from different areas of Azarbaijan, Iran (Fig. 1) and one breeding population (Ghareh Yonjeh) were employed in this study (Table 1). In each population, 30 seeds were randomly selected for individual plant analysis. A mixture of 30 randomly selected seeds per population was also used to prepare bulked DNA sample.

Table 1. List of alfalfa populations used in the present study.

Population number	Population name	Collection site	Elevation (km)	Planting type
1	Gran chay	Kaleibar	750	Irrigated farming, native ecotype
2	Zonorag	Marand	1850	Dry farming, native ecotype
3	Sivan	Marand	2000	Irrigated farming, native ecotype
4	Almalou	Ajabshir	2000	Dry farming, native ecotype
5	Seviar	Hashtrud	1700	Irrigated farming, native ecotype
6	Balsin	Mianeh	1730	Semi-Dry farming, native ecotype
7	Ein-aldin	Bostanabad	1900	Irrigated farming, native ecotype
8	Ilan-jough	Ardabil	1800	Irrigated farming, native ecotype
9	Kordlou	Ahar	1350	Irrigated farming, native ecotype
10	GhareYonje	Khosroshahr	1345	Dry farming, Improved cultivar

DNA isolation: Genomic DNA from 30 individual seeds of each population was extracted following Madden (2002) with mirror modification. The quantity and purity of extracted DNAs were estimated by spectrophotometry and 1% agarose gel electrophoresis. Each DNA sample was diluted to 30 ng and kept at -20°C to use for PCR amplification. DNA from bulked seeds per population was also extracted and referred to as bulked DNA sample.

RAPD amplification: Thirty eight random primers were tested and finally 10 primers were selected in this study for RAPD analysis (Table 2). PCR reactions were performed in a 25 µl total volume containing 1 µl of template DNA (30 ng), 4 pmol of random primers (CinnaGen), 13 µl of 1 X PCR Master Kit (CinnaGen PCR Master Kit, Cat. No. PR8250C) and 10 µl of double distilled H_2O. Amplifications were carried out in a Thermal cycler (Primus 96), programmed for an initial denaturation step at 94°C for 5 min followed by 40 cycles of 1 min at 93°C, 1 min at 40°C, 90 s at 72°C and a final extension cycle of 5 min at 72°C. RAPD products were separated by electrophoresis on 1.5% agarose gels, stained with ethidium bromide, visualized with UV light and then photographed. A 1Kb DNA Ladder (Fermentas) was also loaded to estimate the size of RAPD fragments.

Data analysis: The presence or absence of bands visualized on the gel were scored as 1 (presence) or 0 (absence) for each locus separately. The percentage of polymorphic bands per primer was defined and then within population polymorphism, genetic diversity based on Nei's gene diversity (Nei, 1973) and Shannon's Information index (Lewontin, 1972) and the genetic distances among populations (Nei, 1972) were measured by POPGEN ver 1.32 (Yeh *et al.,* 1999) software. A matrix of pairwise genetic distances was employed to cluster the populations and UPGMA dendrogram was drawn using the sequential agglomerative hierarchical nested (SAHN; Sneath and Sokal, 1973) clustering method as available in NTSYS-pc 2.02 (Rohlf, 1998). Cophenetic correlation was measured with NTYSYS to test the association between input and output of the distance matrix (Mantel, 1967).

Table 2. Properties of arbitrary oligonucleotide primers used for RAPD analysis.

Primers	Sequence (5'-3')	Individual analysis		Bulk analysis	
		Number of polymorphic bands	% of polymorphic bands	Number of polymorphic bands	% of polymorphic bands
OPJ$_4$	CCGAACACGG	19	100.00	10	83.33
B$_1$	GGTTCGCTCC	18	100.00	3	25.00
B$_6$	TGCTCTGCCC	12	100.00	7	63.63
B$_7$	GGTGACGCAG	12	92.31	2	16.67
B$_8$	GTCCACACGG	11	91.67	3	42.86
OPJ$_{13}$	CCACACTACC	20	86.96	5	41.67
B$_{10}$	CTGCTGGGAC	16	88.89	4	28.57
OPA$_1$	CAGGCCCTTC	20	83.33	12	80.00
OPJ$_{19}$	GGACACCACT	15	93.75	3	30.00
OPJ$_{20}$	AAGCGGCCTC	13	86.67	0	00.00
Mean		15.6	92.36	46	41.17

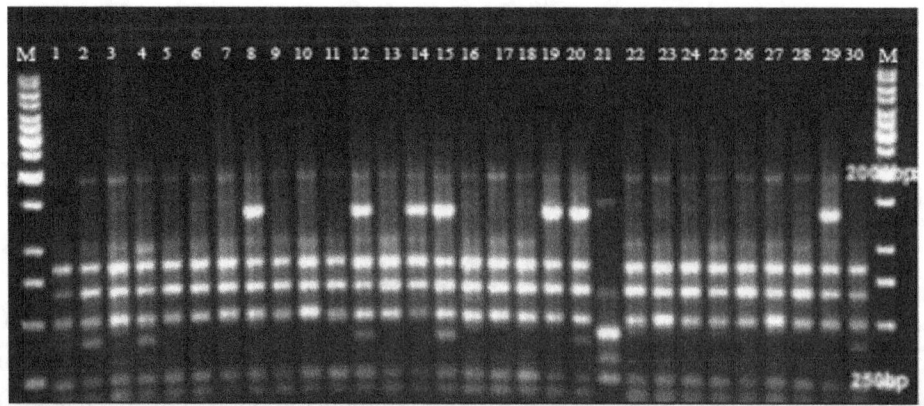

Fig. 2. RAPD fragments for Bostanabad population using the primer B$_6$ in individual analysis. M. Molecular size marker (1 Kb). 1-30. Individuals number.

Analysis of molecular variance (AMOVA) was performed to estimate hierarchical variance components (among individuals within populations, among populations and among groups). AMOVA was carried out via ARLEQUIN 3 (Excoffier *et al.*, 2005). To show a graphical representation of the relationships among populations, principal coordinates analysis (PCoA) was performed using NTSYS-pc, version 2.02. Genetic distances among populations for bulk analysis were estimated and cluster analysis and principal coordinates analyses were performed.

Results

Among 38 random primers tested in this study, 10 primers generated reproducible bands (Table 2). Fig. 2 and Fig. 3 show RAPD fragments in individual plant study and bulk analysis, respectively.

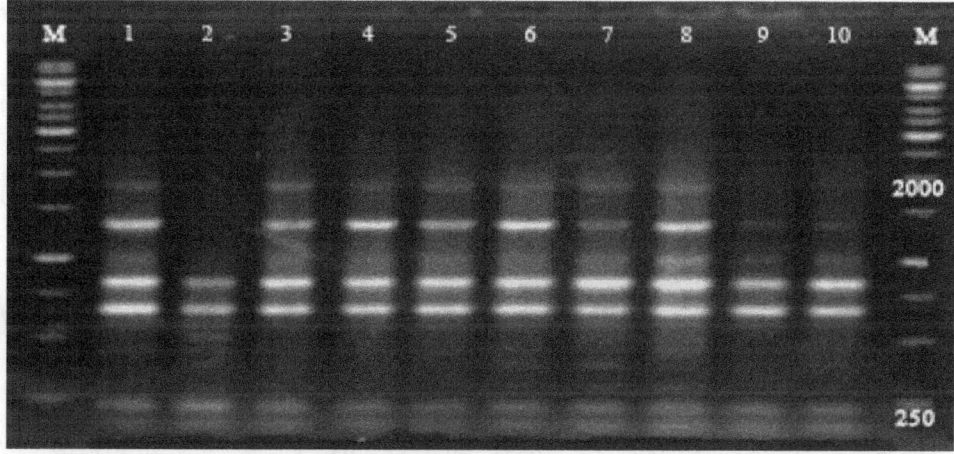

Fig. 3. RAPD fragments for 10 populations using the primer B_6 in bulk analysis. M. Molecular size marker (1 Kb). 1-10. Populations number.

Table 3. Within-population polymorphism and gene diversity ([1]Nei'gene diversity, [2]Shannons information index).

Population number	Number of polymorphic bands	% of polymorphic bands	h[1]	I[2]
1	112	65.88	0.1977	0.3058
2	98	57.65	0.1715	0.2671
3	107	62.94	0.1891	0.2925
4	119	70.00	0.1975	0.3091
5	116	68.24	0.1991	0.3090
6	110	64.71	0.1801	0.2819
7	107	62.94	0.1753	0.2748
8	113	66.47	0.1843	0.2874
9	118	69.41	0.2114	0.3238
10	105	61.76	0.1864	0.2876
Mean	110.5	65	0.1892	0.2939

Individual analysis:

A total of 156 polymorphic bands ranging from 250 to 2500 bp were identified. Three primers (B_1, B_6 and OPJ_4) produced 100% polymorphic bands. Minimum percentage of polymorphic bands was observed by primer OPA_1 (Table 2). The percentage of polymorphic bands within populations differed from 57.65% for population 2 to 70% for population 4 (Table 3). Additionally, the populations 9 and 2 showed the maximum and minimum genetic diversity (Table 3), respectively. Total genetic diversity (H_T) and within population genetic diversity (Hs) were calculated as 0.2349 and 0.1892, respectively and the degree of genetic differentiation among populations (G_{ST}) was estimated as 0.1944. These results indicated that diversity within populations was greater than that among populations. Genetic distances among pairs of populations ranged from 0.025 between populations 7 and 8 to 0.1103 between populations 2 and 10. The average distance among populations was 0.0631. In total, genetic distances among populations were low (Table 4).

Table 4. Nei's genetic distances between populations for individual analysis (lower diagonal) and bulk analysis (upper diagonal). Mean for upper diagonal: 0.1169; Mean for lower diagonal: 0.0631

Population	1	2	3	4	5	6	7	8	9	10
1		0.0919	0.0543	0.0857	0.1149	0.0703	0.0870	0.1243	0.1363	0.0869
2	0.0341		0.1150	0.0667	0.1220	0.1542	0.1149	0.1428	0.1566	0.1264
3	0.0391	0.0417		0.0857	0.0919	0.0595	0.0869	0.1135	0.1136	0.0543
4	0.0616	0.0859	0.0595		0.0667	0.1250	0.1200	0.1477	0.1617	0.1200
5	0.0490	0.0779	0.0652	0.0413		0.1314	0.1149	0.1314	0.1253	0.1149
6	0.0487	0.0528	0.0447	0.0575	0.0428		0.1027	0.1290	0.1751	0.0919
7	0.0358	0.0438	0.0392	0.0774	0.0620	0.0538		0.0702	0.1023	0.0543
8	0.0434	0.0701	0.0506	0.0553	0.0491	0.0530	0.0250		0.1073	0.1027
9	0.0760	0.0886	0.0830	0.0978	0.0934	0.0848	0.0792	0.0645		0.0795
10	0.0674	0.1103	0.0978	0.1050	0.0831	0.0950	0.0872	0.0663	0.0791	

Genetic distance values were used to construct a UPGMA dendrogram and populations were divided into three groups (Fig. 4). First groups included population 10 (a breeding population) and second group included population 9. Other populations belonged to third group. Matrix correlation was estimated as 0.849. To study relationships among populations, AMOVA was performed based on population clustering (significance tests were provided by computing 1023 permutations). Significant differences were observed among groups, among populations within groups and among individuals within populations. However, the high genetic variation (76.08 %) was attributed to differences within populations (Table 5).

AMOVA was also performed in population level to estimate diversity within and between populations ($F_{ST} = 0.186$; p = 0.05). Although variation among populations was significant, the great genetic diversity (81.37%) was observed within populations (Table 5). Fig. 5 shows the results of PCoA. On the basis of the first and second coordinates, which accounted for 29.29% and 19.29% of the total variation, respectively, populations were distributed in three groups. Populations 10 and 9 belonged to first and second groups, respectively and the other populations belonged to third group.

Table 5. Results of analysis of molecular variance (AMOVA) in individual analysis.

Source of variation	df	SS	Variance component	Percentage of variation	P
Based on clustering					
Among groups	2	453.744	2.34129	10.46	0.023
Among populations within groups	7	752.262	3.01445	13.46	$<10^{-5}$
Within populations	290	4939.467	17.03264	76.08	$<10^{-5}$
Total	299	6145.473	22.38839		
In population level					
Within populations	290	4939.467	17.03264	81.37	$<10^{-5}$
Among populations	9	1206.007	3.89894	18.63	$<10^{-5}$
Total	299	6145.473	20.93158		
F_{ST}			0.18627		

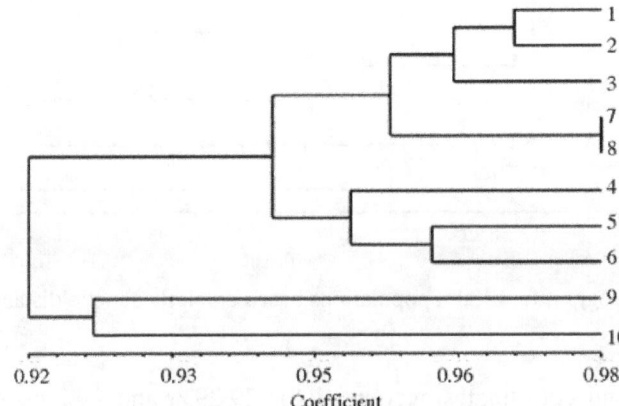

Fig. 4. UPGMA dendrogram for alfalfa populations based on Nei's genetic distances in individual analysis.

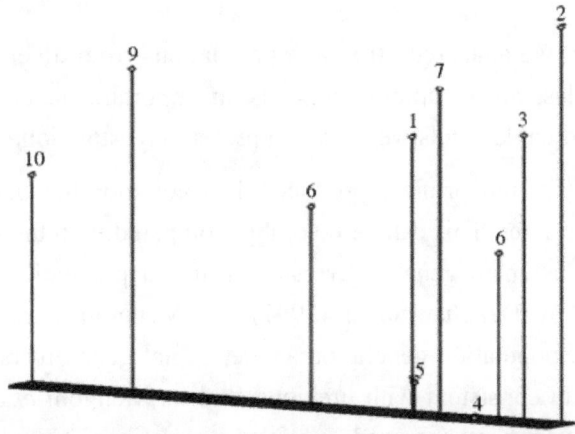

Fig. 5. Principal coordinates analysis (PCoA) for alfalfa populations based on the first and second coordinates (in individual analysis).

Bulk analysis:

A total 46 polymorphic bands were identified in bulk analysis. Maximum and minimum percentages of polymorphic bands were observed by primers OPJ_4 and OPJ_{20}, respectively. The average percentage of polymorphic bands was 41.17 % (Table 2). Cluster analysis based on Nei's genetic distances (Table 4) divided populations into three groups (Fig. 6). First group included populations 8 and 9, the second groups included populations 5, 4, and 2 and third groups included the others. Matrix correlation was estimated as 0.712. PCoA was performed for bulk samples and populations were located into 3 groups (Fig. 7).

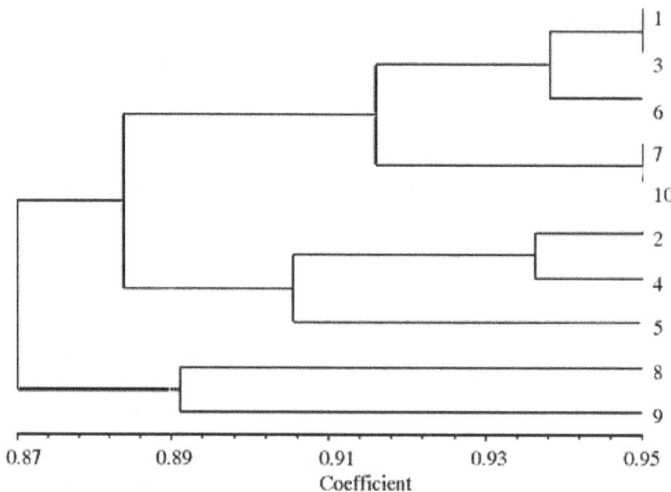

Fig. 6. UPGMA dendrogram for alfalfa populations based on Nei's genetic distances in bulk analysis.

The first and second coordinates accounted for 29.29% and 19.29% of the total variation, respectively. In total, results of bulk analysis were different from results of individual analysis.

Discussion

In the present study we analyzed 10 alfalfa populations from diverse regions of Azarbaijan, Iran using RAPD profiles. Since reproducibility is an important factor in RAPD studies (Ulloa *et al.*, 2003), only reproducible bands were used in present investigation.

In individual analysis, ten primers produced 156 polymorphic bands with an average 15 polymorphic bands per primer. This can be favorably compared with the number of bands used by Tucak *et al.* (2008) to estimate genetic diversity in alfalfa populations and is higher than the number of bands used by Dehghan-Shoar (1997) and Mengoni *et al.* (2000) to study alfalfa populations. In terms of population genetic parameters, total gene diversity (H_T) observed in this study was high. It was in consistent with previous studies. Mengoni *et al.* (2000) suggested that high level of genetic diversity is observed in alfalfa populations. Moreover, Falahati-Anbaran *et al.* (2007) studied population genetic structure in alfalfa from various regions contiguous to the centers of origin of the species. They proposed that since northwestern of Iran is the primary centre of diversity for alfalfa, so high level of genetic diversity exists within and among Iranian alfalfa populations such as populations employed in this work. However, the within population diversity (based on Nei's gene diversity) was high for each population, as found in previous studies (Flajoulot *et al.*, 2005; Falahati-Anbaran *et al.*, 2007; Tucak *et al.*, 2008). Gherardi *et al.* (1998) also suggested that the within population diversity is higher than diversity among populations. It can be explained by the outcrossing and tetraploid nature of alfalfa that results in highly heterogeneous and heterozygous populations (Kidwell *et al.*, 1994).

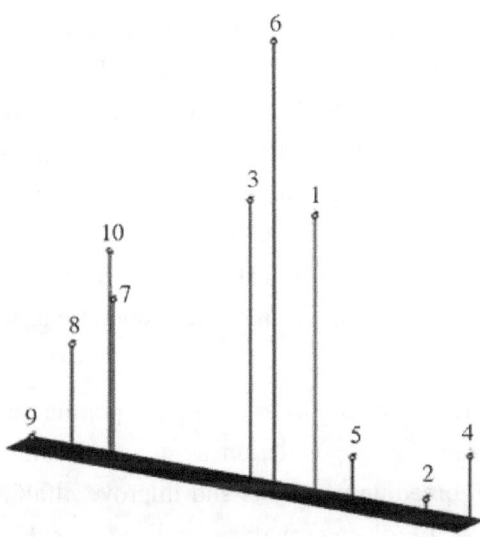

Fig. 7. Principal coordinates analysis (PCoA) for alfalfa populations based on the first and second coordinates in bulk analysis.

Results of analysis of molecular variance suggested that the largest proportion of genetic variation was attributed to variation among individuals within populations (81.37%). These results were in agreement with previous studies (Falahati-Anbaran *et al.*, 2007; Tucak *et al.*, 2008). Low genetic distances were detected between populations possibly due to small geographical distances existed between them. In spite of it, cluster analysis could group populations and AMOVA based on population grouping showed a significant distance between groups. The largest genetic distance was observed between populations number 10 (Ghareh Yonjeh) and number 2 (Zonorag). Separation of Ghareh Yonjeh which is a breeding population from the other populations indicates the sufficiency of this method to study relationship in alfalfa populations. Falahati-Anbaran *et al.* (2007) could also separate Ghareh Yonjeh from other Iranian alfalfa populations. Population number 9 was clustered into a distinct group. Morphological studies indicated differences among this population and other populations. Thereupon, separation of it from other populations of Azarbaijan can be related to morphological differences. Other populations grouped together in one cluster and formed a different branch in the dendrogram. Distribution of these populations on distinct branch was in agreement with geographical patterns of them.

Cluster analysis in bulk method could not separate Ghareh Yonjeh from other populations and genetic differentiation of populations was not in agreement with geographical or morphological patterns. Such differences among results obtained from bulk analysis and individual analysis were observed in previous studies (Mengoni *et al.*, 2000; Pupilli *et al.*, 2000). Negri *et al.* (1995) and Pupilli *et al.* (2000) reported that bulk procedure reduces within population diversity when frequency of polymorphic fragments is low. Some DNA sequences are found in a few individuals and produce rare fragments in individual analysis. Since these sequences compose a low concentration of template DNA in bulked sample, they can not efficiently be amplified. So rare fragments observed in individual analysis are absent in bulk analysis (Yu and Pauls, 1993) as observed in our study. Kidwell *et al.* (1994) also proposed that bulk method underestimates the level of genetic diversity in both within and between populations and results of differentiation among populations in bulk analysis are not in agreement with individual analysis. This was true especially in our study with 30 individuals per bulk sample. Since using of greater number of individuals in bulk samples reduces the probability of detecting rare fragments that may be diagnostic of a population, so differentiation between populations was not carried out precisely (Kidwell *et al.*, 1994).

The high level of genetic diversity observed within populations in our study, particularly population 4, 5 and 9, indicates each population as a genetic source for selection of suitable genotypes to employ them in breeding programs and improve alfalfa cultivars with high level of heterosis. Furthermore results of cluster analysis indicated that GhareYonje and population number 9, are different from other populations. Thus each of them can be used as parents in breeding programs. Bulk method offers a rapid analysis of RAPD patterns in genetic study of alfalfa population. However, comparison of bulk analysis with individual analysis showed that it is better to use individual analysis in detection of relationships among alfalfa populations and estimation of genetic diversity especially within populations.

Finally, RAPD analysis was demonstrated as a suitable method to study genetic diversity and relationships among alfalfa populations. However it is advised to accompany results of RAPD procedure with other molecular methods and morphological studies.

Acknowledgments

This research was performed in the Agricultural Biotechnology Research Institute of Iran (ABRII), for Northwest and West of Iran, Tabriz-Ian. We thank Ms Nahid Hosseinzadeh for her editorial assistance.

References

Arzani, A. and Samei, K. 2004. Assessment of genetic diversity among Persian clover cultivars as revealed by RAPD markers. *In*: Vollmann, J., Grausgruber, H. and Ruckenbauer, P. (ed.), Genetic variation for plant breeding. EUCARPIA and BOKU, University of Natural Resources and Applied Life Sciences, Vienna, pp. 85-88.

Dehghan-Shoar, M., Hampton, J.G. and Gardiner, S.E. 1997. Genetic analysis among and within populations forming ecotypes and cultivars of lucerne, *Medicago sativa* (Leguminosae), using RAPD fragments. Pl. Syst. Evol. **208**: 107-119.

Excoffier, L., Laval, G. and Schneider, S. 2005. Arlequin (version 3.0): An integrated software package for population genetics data analysis. Evol. Bioinf. Online **1**: 47-50.

Falahati-Anbaran, M., Habashi, A.A., Esfahany, M., Mohammadi, S.A. and Gharayazie, B. 2007. Population genetic structure based on SSR markers in alfalfa (*Medicago sativa* L.) from various regions contiguous to the centres of origin of the species. J. Genet. **86**: 59-63.

Flajoulot, S., Ronfort, J. Baudouin, P., Barre, P., Huguet, T., Huyghe, C. and Julier, B. 2005. Genetic diversity among alfalfa (*Medicago sativa)* cultivars coming from a breeding program, using SSR markers. Theor. Appl. Genet. **111**: 1420-1429.

Gherardi, M., Mangin, B., Goffinet, B., Bonnet, D. and Huguet, T. 1998. A method to measure genetic distance between allogamous populations of alfalfa (*Medicago sativa*) using RAPD molecular. Theor. Appl. Genet. **96**: 406-412.

Hanson, A.A., Barnes, D.K. and Hill, R.R. 1988. Alfalfa and alfalfa improvement. Madison, Wisconsin, USA.

Jenczewski, E., Prosperi, J.M. and Ronfort, J. 1999. Differentiation between nature and cultivated population of *Medicago sativa* (Leguminosae) from Spain. Analysis with random amplified polymorphic DNA (RAPD) markers and comparison with allozymes. Mol. Ecol. **8**: 1317-1330.

Kidwell, K.K., Austin, D.F. and Osborn, T.C. 1994. RFLP evaluation of nine Medicago accessions representing the original germplasm sources of North American alfalfa cultivars. Crop Sci. **34**: 230-236.

Lewontin, R.C. 1972. The apportionment of human diversity. Evol. Biol. **6**: 38-398.

Madden, D. 2002. Investigating plant DNA. Technical guide. NCBE. The University of Reading, United Kingdom.

Mantel, N.A. 1967. The detection of disease clustering and generalized regression approach. Cancer Res. **27**: 209-220.

Mengoni, A., Gori, A. and Bazzicalupo, M. 2000. Use of RAPD and microsatellite (SSR) variation to assess genetic relationships among populations of tetraploid alfalfa, *Medicago sativa*. Plant Breed. **119**: 311-317.

Michelmore, R.W., Paran, I. and Kesseli, R.V. 1991. Identification of markers linked to disease-resistance genes by bulked segregate analysis: A rapid method to detection markers in specific genomic regions by using segregating populations. Proc. Natl. Acad. Sci. USA. **88**: 9828-9832.

Negri, V., Barcaccia, G., Russi, L., Tavoletti, S., Pellicoro, A. and Falcinelli, M. 1995. RAPD fingerprinting as a tool for characterizing the genetic background of Lucerne (*Medicago sativa* L.) landraces. Universita degli Studi di Ancona, Italy.

Nei, M. 1972. Genetic distance between populations. Am. Nat. **106**: 283-292.

Nei, M. 1973. Analysis of gene diversity in subdivided population. Proc. Natl. Acad. Sci. USA. **70**: 3321-3323.

Pupilli, F., Labombarda, P., Scott, C. and Arcioni, S. 2000. RFLP analysis allows for the identification of alfalfa ecotypes. Plant Breed. **119**: 271-276.

Rahman, M.O. 2006. Evaluation of RAPD markers for taxonomic relationships in some aquatic species of *Utricularia* L. (*Lentibulariaceae*). Bangladesh J. Plant Taxon. **13**(2): 73-82.

Rohlf, F.J. 1998. NTSYS-pc, Numerical taxonomy and multivariate analysis system, version 2.02. Exeter Software, New York, USA.

Rowe, D.E. and Hill, R.R. 1999. Breeding theory and the development of alfalfa. The Alfalfa Genome. (www.naaic.org/TAG/TAGpapers/RoweAbs.html).

Sneath, P.H.A. and Sokal, R.R. 1973. Numerical Taxonomy. Freeman, San Francisco.

Tucak, M., Popovic, S., Cupic, T., Bolaric, S. and Kozumplic, V. 2008. Genetic diversity of alfalfa (*Medicago* spp.) estimated by molecular markers and morphological characters. Period. Biol. **110**(3): 243-249.

Ulloa, O., Ortega, F. and Campos, H. 2003. Analysis of genetic diversity in red clover (*Trifolium pratense* L.) breeding populations as revealed by RAPD genetic markers. Genome 46(4): 529-535.

Vandemark, G.J., Ariss, J.J., Bauchan, G.A., Larsen, C.R., Hughes, J.T. 2006. Estimating genetic relationships among historical sources of alfalfa germplasm and selected cultivars with sequence related amplified polymorphisms. Euphytica **152**: 9-16.

Veronesi, F., Brummer, E.C. and Huyghe, C. 2010. Alfalfa. *In:* Boller, B., Posselt, U.K. and Veronesi, F. (ed.), Fodder Crops and Amenity Grasses. Series: Handbook of Plant Breeding, Springer, New York, USA. **5**: 395-437.

Volenec, J.J., Cunningham, S.M., Haagenson, D.M., Berg, W.K., Joern, B.C. and Wiersma, D.W. 2002. Physiological genetics of alfalfa improvement: past failures, future prospects. Field Crop Res. **75**: 97-110.

Williams, J.G.K., Kubelik, A.R., Livak, K.J., Rafalski, J.A. and Tingey, S.V. 1990. DNA polymorphisms amplified by arbitrary primers are useful as genetic markers. Nucleic Acids Res. **18**: 6531-6535.

Yeh, F.C., Yang, R.C., Boyle, T.B.J., Ye, Z.H. and Mao, J.X. 1999. POPGENE (the user-friendly shareware for population genetic analysis), version 1.32. Molecular Biology and Biotechnology Centre, University of Alberta, Edmonton, Canada.

Yu, K. and Pauls, K.P. 1993. Rapid estimation of genetic relatedness among heterogeneous populations of alfalfa by random amplification of bulked genomic DNA samples. Theor. Appl. Genet. **86**: 788-794.

17

PHYLOGENY OF *GALIUM* L. (RUBIACEAE) FROM KOREA AND JAPAN BASED ON CHLOROPLAST DNA SEQUENCE

KEUM SEON JEONG, JAE KWON SHIN[1], MASAYUKI MAKI[2] AND JAE-HONG PAK[3]

Division of Forest Biodiversity, Korea National Arboretum, Pocheon, Gyeonggi-do 487-821, Korea

Keywords: Chloroplast DNA; Korean-Japan *Galium;* Molecular data; Phylogeny.

Abstract

The present paper deals with the phylogeny and inter-and intragenic relationships using four chloroplast DNA sequences within 19 *Galium* L. species from Korea and Japan. Maximum parsimony and Bayesian analyses were conducted to clarify the relationships among the section and species. The strict consensus tree had three main clades. Clade I comprises of the only individuals of *G. paradoxum* Maximowicz (sect. *Cymogalia*), which is distinguished by opposite leaves in the genus, supported by the 100% bootstrap value (PP: 0.98); Clade II consists of members of eight sections (sect. *Galium*, sect. *Hylaea*, sect. *Kolgyda*, sect. *Trachygalium*, sect. *Leptogalium*, sect. *Orientigalium*, sect. *Aparine*, and sect. *Leiogalium*); Clade III comprises members of eight sections (sect. *Baccogalium*, sect. *Lophogalium*, sect. *Platygalium*, sect. *Relbunium*, sect. *Depauperata*, sect. *Aparinoides,* sect. *Leiogalium* and *Trachygalium*). The sect. *Leptogalium* which includes two taxa namely *G. tokyoense* Makino and *G. dahuricum* var. *lasiocarpum* (Makino) Nakai is paraphyletic. Four taxa of *Trachygalium* group (*G. trachyspermum* A. Gray, *G. gracilens* (A. Gray) Makino, *G. pogonanthum* Franch. & Sav., *G. koreanum* Nakai) were placed from sect. *Cymogalia* to sect. *Platygalium* based on molecular and morphological data.

Introduction

Galium L., the largest genus of the tribe Rubieae in the family Rubiaceae (Robbrecht and Manen, 2006), is taxonomically diverse and comprises over 650 species (Govaerts, 2006). *Galium* is divided into 16 sections based on characters of leaf and fruit by Ehrendorfer *et al.* (2005). The species of *Galium* are distributed centrally in temperate regions and are mostly annual and perennial herbaceous plants. The genus is characterized by more than two leaf-like whorls, number of divided petal, rudimentary calyx and a two locular ovary.

Phylogenetic relationships among species of tribe Rubieae including eleven genera have been studied by many researchers (Ehrendorfer *et al.*, 1994, 2014; Manen *et al.*, 1994; Manen and Natali, 1995; Natali *et al.*, 1995, 1996; Soza and Olmstead, 2010). Molecular phylogenetic studies using chloroplast DNA *atp*B-*rbc*L intergenic region have shown monophyly of the tribe Rubieae with seven major clades, and confirmed that genera *Asperula* and *Galium* is not a monophyletic group (Manen *et al.*, 1994; Natali *et al.,* 1995, 1996). Soza and Olmstead (2010) conducted more clearly molecular phylogenetic analysis of tribe Rubieae using three chloroplast DNA makers and their results indicated that *Galium* is polyphyletic, and species of *Galium* occur in three major clades (Clades III, V, VII). Recently, phylogenetic relationships study of tribe Rubieae including

[1] Division of Forest Resource Conservation, Korea National Arboretum, Pocheon, Gyeonggi-do 487-821, Korea.
[2] Division of Ecology and Evolutionary Biology, Graduate School of Life Sciences, Tohoku University, Aoba, Sendai 980-8578, Japan.
[3] Research Institute for Dok-do and Ulleung-do Island, Kyungpook National University, Daegu 702-701, Korea. Corresponding author. Email: jhpak@knu.ac.kr

some *Galium* species by Ehrendorfer *et al*. (2014) has evaluated that genus *Galium* is paraphyletic. Although there have been several phylogenetic study to investigate relationships of tribe Rubieae, very little is known about phylogenetic relationships among Korean species of *Galium*. Soza and Olmstead (2010) determined the phylogenetic relationships among Rubieae including members of *Galium* but this study included only three common species distributed in Korea and Japan. In Korea, twenty taxa of seven sections are currently recognized (Lee, 1995; Lee, 1979; Lee, 2004). *G. koreanum* Nakai, *G. verum* var. *asiaticum* for. *pusillum* (Nakai) M. Park are endemic to Korea and latter species is restrictedly distributed in Mt. Halla of Jeju Island. *G. kikumugura* Ohwi is broadly expanded to Japan. Jeong and Pak (2009, 2012) conducted morphological and somatic chromosome number counts of Korean *Galium*. These studies however, provided very little phylogenetic relationships among the species. Therefore, further studies are needed to understand their phylogenetic relationships among Korean *Galium* species and taxonomic position of Korean and Japan taxa within the *Galium* spp. occurring worldwide. This study aims to clarify inter-and intragenic relationships within Korean and 10 Japanese *Galium* species, and to determine the taxonomic position of Korean endemic taxa within the closely related *Galium* spp. using the chloroplast DNA sequences.

Materials and Methods

Plant materials

Total 19 species of *Galium* distributed in Korea and Japan were collected (Table 1). We selected two outgroup taxa [*Didymaea alsinoides* (Cham. and Schltdl.) Standl., and *Rubia cordifolia* L.] based on the results of the analyses of Soza and Olmstead (2010). The sequences of *Galium* and outgroups obtained from National Center for Biotechnology Information (NCBI) database with the exception of sequences of sample from Korea-Japan. All sources and voucher specimens of materials were deposited at the Herbarium of Kyungpook National University (KNU).

DNA extraction, amplification and sequencing

Total genomic DNA was extracted from fresh leaf tissues and field-collected silica-gel dries tissue using the 2 % hexa decyltrimethyl ammonium bromide (CTAB) procedure (Doyle and Doyle, 1987). We amplified the *rpo*B-*trn*C region and *trn*C-*ycf*6 region with primers designed by Demesure *et al*. (1995). The *trn*L-*trn*F-*ndh*J region was amplified using primers published in Taberlet *et al*. (1991) and Shaw *et al*. (2007) (Table 2). Polymerase chain reaction (PCR) conditions were an initial denaturation of 94°C for 5 min, 35 cycles of 94°C denaturation for 30 s, 48°C-57°C annealing for 30 s extension for 1m, and final extension at 72°C for 10 min. PCR products were purified using the QIAquick PCR purification kit following the instructions of the manufacturer. Sequencing reactions were carried out for the purified PCR products using Big Dye Terminator Cycle Sequencing reagents (Applied Biosystem, Foster city, CA, USA). For sequencing, we used the same primers as those used for PCR. All sequences have been deposited in GenBank (Table 1).

Data analysis

The DNA sequences were aligned with Clustal X (Thompson *et al*., 1997). All chloroplast regions were combined and analyzed using Maximum Parsimony (MP) and the Bayesian analyses. Gaps introduced from the alignment were treated as missing characters in subsequent analyses. MP analyses were conducted in a PAUP* (version 4.0b 10; Swofford, 2003) using a heuristic searches with TBR branch swapping and MULTREES option. Relative support of various monophyletic groups revealed in the most parsimonious trees was examined with the bootstrap

Table 1. Sampling sites of plant materials used for phylogenetic analyses.

Taxon	Locality	Voucher	GenBamk acc. No.			
			trnC-ycf6	trnF-ndhJ	TrnL	rpoB-trnC
Sect. *Aparine*						
Galium spurium var. *echinospermon*	Chilgok-gun, Korea	J20050310	KC339150	KC339020	KC339085	LC062539
Sect. *Aparinoides*						
G. trifidum	Jeju-si, Korea	J20060807	KC339148	KC339018	KC339083	LC062537
	Tokyo metro, Japan	M20100501	KC339149	KC339019	KC339084	LC062538
Sect. *Cymogalia*						
G. paradoxum	Pyeongchang-gun, Korea	J20090814	KC339164	KC339034	KC339099	LC062552
	Jeongseon-gun, Korea	J20050618	KC339163	KC339033	KC339098	LC062551
	Muju-gun, Korea	J20100844	KC339162	KC339032	KC339097	LC062550
Sect. *Hylaea*						
G. trifloriforme	Ulleung-gun, Nari, Korea	J20080621	KC339204	KC339074	KC339139	LC062581
	Ulleung-gun, Korea	J20080635	KC339203	KC339073	KC339138	LC062580
	Ulleung-gun, Taehwa, Korea	J20080603	KC339205	KC339075	KC339140	LC062582
	Miyagi, Japan	J20100748	KC339206	KC339076	KC339141	LC062583
G. japonicum	Ulleung-gun, Nari, Korea	J20080611	KC339207	KC339077	KC339142	LC062584
	Ulleung-gun, Nari, Korea	J20080612	KC339151	KC339021	KC339086	LC062540
	Jeongeup-si, Korea	J20100845	KC339209	KC339079	KC339144	LC062585
	Jeju-si, Korea	J20070901	KC339210	KC339080	KC339145	LC062586
	Miyagi, Japan	J20100758	KC339211	KC339081	KC339146	LC062587
	Yamagata, Japan	J20100759	KC339212	KC339082	KC339147	LC062588
Sect. *Leptogalium*						
G. dahuricum var. *lasiocarpum*	Namyangju-si, Korea	J20100897	KC339189	KC339059	KC339124	LC062569
	Pyeongchang-gun, Korea	J20090807	KC339192	KC339062	KC339127	LC062571
	Yeongwol-gun, Korea	J20080926	KC339155	KC339025	KC339090	LC062543
	Seongju-si, Korea	J20100658	KC339188	KC339058	KC339123	LC062568
	Jecheon-si, Korea	J20091021	KC339190	KC339060	KC339125	LC062570
	Yamagata, Japan	J20100708	KC339194	KC339064	KC339129	LC062573
G. kikumugura	Mt. Zao, Japan	J20100765	KC339200	KC339070	KC339135	LC062577
G. pseudoasprellum	Miyagi, Japan	J20100789	KC339202	KC339072	KC339137	LC062579
G. tokyoense	Pocheon-si, Korea	J20070938	KC339195	KC339065	KC339130	LC062572
	Pocheon-si, Korea	J20090808	KC339193	KC339063	KC339128	LC062574
	Tokyo metro, Japan.	M20090503	KC339197	KC339067	KC339132	LC062575
Sect. *Platygalium*						
G. boreale	Yeongwol-gun, Korea	J20050625	KC339152	KC339022	KC339087	LC062541
	Mongolia	L20090830	KC339153	KC339023	KC339088	LC062542
G. gracilens	Sunchen-si, Korea	J20090801	KC339181	KC339051	KC339116	LC062566
	Hwasun-gun, Korea	J20090830	KC339180	KC339050	KC339115	LC062565
G. kamtschaticum var. *yakusimense*	Jeju-si, Korea	J20070907	KC339166	KC339036	KC339101	LC062553
G. koreanum	Sancheong-gun, Korea	J20100808	KC339186	KC339056	KC339121	LC062567
G. kinuta	Yeongwol-gun, Korea	J20050626	KC339167	KC339037	KC339102	LC062554
G. pogonanthum	Hamyang-gun, Korea	J20090504	KC339172	KC339042	KC339107	LC062559
	Jeju-si, Korea	J20050706	KC339171	KC339041	KC339106	LC062558
G. trachyspermum	Inje-gun, Korea	J20080906	KC339170	KC339040	KC339105	LC062546
	Andong-si, Korea	J20070751	KC339157	KC339027	KC339092	LC062545
	Gyeongju-si, Korea	J20100913	KC339159	KC339029	KC339094	LC062547
	Geoje-si, Korea	J20090327	KC339156	KC339026	KC339091	LC062544
	Miyagi, Japan	J20100723	KC339160	KC339030	KC339095	LC062548
	Yamagata, Japan	J20100747	KC339161	KC339031	KC339096	LC062549
Sect. *Galium*						
G. verum var. *asiaticum*	Geoje-si, Korea	J20100524	KC339173	KC339043	KC339108	LC062563
	Jeju-si, Korea	J20090685	KC339174	KC339044	KC339109	LC062562
	Fukui, Japan	M20100503	KC339176	KC339046	KC339111	LC062561
G. verum var. *trachycarpum* f. *nikkoense*	Ulsan metro., Korea	J20050830	KC339198	KC339068	KC339133	LC062576
	Tokushima, Japan	J20100732	KC339177	KC339047	KC339112	LC062564
G. verum var. *asiaticum* f. *pusillum*	Jeju-si, Korea	J20050807	KC339175	KC339045	KC339110	LC062560

method (Felsenstein, 1985). Bootstrap values were calculated from 1,000 replicates with the random addition and heuristic search options. The Bayesian phylogenetic analyses were conducted with MrBayesver 3.1.2 (Ronquist and Huelsenbeck, 2003). The suitable model was determined to be GTR+I+G for combined sequence data by MrModeltest 2.3 (Nylander, 2004). Each Morkov chain was started from a random tree and run for 1,000,000 generations, sampling a tree every 100 generations. Burn-in time was estimated from the plot of likelihoods generated using the 'sump' command in MrBayes. Posterior probabilities (pp) were based on analysis of post-burn-in tree. Nodes were considered highly supported when pp values were higher than 0.95 (Felesenstein, 1985).

Results and Discussion

Sequence characteristics

The total of 4,341 lengths of the aligned sequences was used for phylogenetic analysis. Of a total of investigated character sites, 2,793 characters were constant and 824 characters were parsimony informative including out groups. The parsimony analyses generated 10,620most parsimonious trees with a total length of 2,970 steps, a consistency index of 0.65 and a retention index of 0.88.The MP tree with bootstrap values(BP) and PP are shown in Fig. 1.

Phylogenetic analyses

The strict consensus tree had three main clades (clade I, clade II and clade III). Clade IV is highly supported by the 100% bootstrap value (PP: 0.98) and was sister to the rest of the species, which were grouped in two other clades. This clade was only composed of the individuals of *G. paradoxum* Maxim. Clade IIa is supported 99% bootstrap value (PP<0.95). Clade IIb consists of two highly supported subclades (subclade IIa and IIb). Subclade IIa included three taxa: *G. dahuricum* var. *lasiocarpum* (Makino) Nakai., *G. pseudoasprellum* Makino and *G. triflorum* Michx. comprising of Group B. *G. triflorum* (sect. *Trachygalium*) was sister to *G. dahuricum* var. *lasiocarpum* from Korea-Japan and *G. pseudoasprellum* from Japan (99% bootstrap value). Subclade IIb is supported by 91% bootstrap value (PP<0.95). This subclade contained 8 taxa from Korea-Japan. It was further divided into Group C and D. Group C contained members of three sections (sect. *Galium*, sect. *Leiogalium* and sect. *Leptogalium*) which are identified by Soza and Olmstead (2010), *G. tokyoense* Makino, *G. kikumugura*, and three species belonging to sect. *Galium* from Korea-Japan. But the *G. verum* group from Korean and Japanese were not well resolved. In the Group D, *G. japonicum* (Maxim.) Makino & Nakai from Korea and Japan is monophyletic, although the individuals of *G. trifloriforme* Kom. did not form monophyletic group. These two taxa share its most recent common ancestor with *G. spurium* var. *echinospermum* (Wallr.) Hayekand *G. odoratum* (L.) Scop (61% bootstrap value (PP: 0.96)). Clade III is supported by 91% bootstrap value (PP: 0.97), comprising eight sections; sect .*Baccogalium*, sect. *Lophogalium*, sect. *Platygalium*, sect. *Leiogalium*, sect. *Trachygalium*, sect. *Relbunium*, sect. *Depauperata*, sect. *Aparinoides*. The members of sect. *Depauperata*, and sect. *Aparinoides* are sister to the rest of the species within this Clade. *G. trifidum* L. is paraphyletic and unresolved within the clade. Group A in Clade III included four taxa from *G. trachygalium* group (*G. gracilens* (A. Gray) Makino, *G. koreanum*, *G. pogonanthum* Franch. & Sav. and *G. trachyspermum* A. Gray) and members of sect. *Platygalium* (BS: 80%, PP<0.95). The previous classification based on morphological study of the four taxa of the *G. trachygalium* group was not resolved (Jeong and Pak, 2009). The individuals from the same taxa did not even form the monophyletic. *G. kinuta* Nakai & Hara belonging to sect. *Platygalium* with *G. boreale* L. was resolved as paraphyletic.

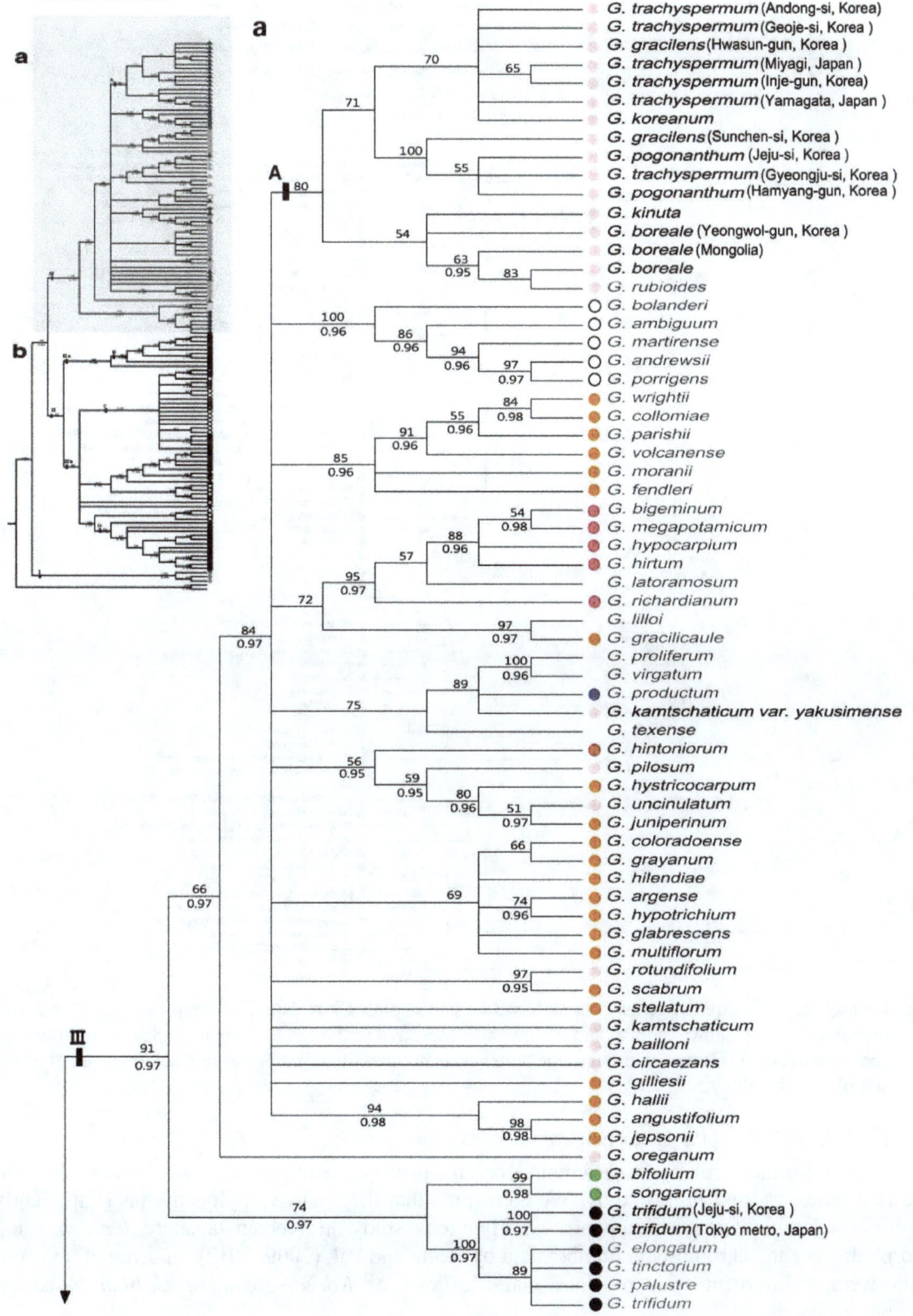

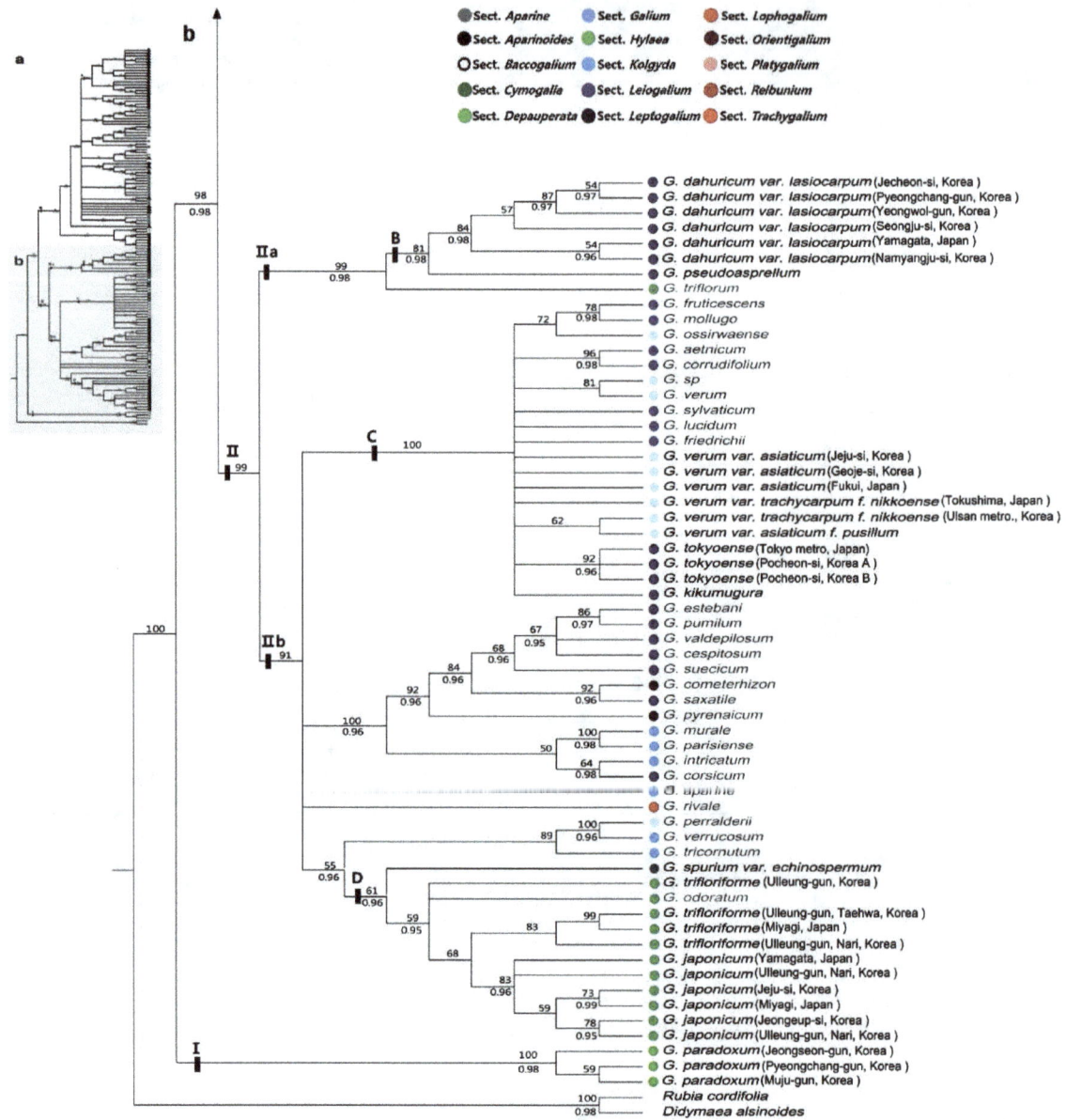

Fig. 1. Strict consensus tree of genus *Galium* based on Chloroplast DNA data, Bootstrap values and posterior probabilities are shown above and below branches, respectively. Different shapes were used for sectional treatments (taxon without shape "represents not classified"). Species in black represent the taxa sampled in this study.

Phylogenetic relationships of Korean-Japanese Galium

The phylogenetic relationships among Korean *Galium* and some of Japanese *Galium* were, for the first time, assessed in this study. We confirmed that the cpDNA phylogeny has significantly higher resolution and better support than previous study in Korean-Japanese *Galium* using morphological and chromosome number data by Jeong and Pak (2009, 2012). In some of taxa, our data were incongruent with previous classifications of *Korean-Japanese Galium* based on morphological data.

G. *paradoxum* was sister to the group consisting of the rest of the *Galium* species (Fig 1). It also support the study of Ehrendorfer *et al*. (2014) using the plastid DNA sequences. The species is a perennial herb with opposite leaves, a pair of scale-like small stipules, one vein, white petiole and corolla, and rotate flowers. *G. paradoxum* was placed into a sect. *Cymogalia* based on the characters of inflorescence and hairs of a fruit (Pobedimova *et al*., 2000; Ehrendorfer *et al*., 2005). Its main distributions is in eastern Asia (Ehrendorfer *et al*., 2014), and mainly occurs in moist high elevations in mountain forests.

The taxa in the clade II have whorls of six or eight leaf-like organs. The five taxa from Korea-Japan are contained in Group C. The taxa of *G.verum* group (sect. *Galium*; *G. verum* var. *asiaticum* Nakai, *G. verum* f. *nikkoense* var. *trachycarpum* (Nakai) Ohwi and *G. verum* var. *asiaticum* f. *pusillum*) showed polytomies in the MPtree with weak PP. *G. verum* var. *asiaticum* is widely distributed throughout Korea and Japan. In our study, G. *verum* var. *asiaticum* have five chloroplast types from five individuals. But we cannot find morphological variation among the individuals. The three taxa are erect and have whorls of six or more than leaf-like organs, inflorescences of branched panicles with white or yellow flowers, and glabrous fruits. These three taxa don't exhibit significant morphological differences. But the plant and leaves size of *G. verum* var. *asiaticum* f. *pusillum* are smaller than those of other two taxa, and Korean endemic species in Mt. Halla on Jeju Island (Lee, 2004). It formed a clade with *G. verum* f. *nikkoense* var. *trachycarpum* from Ullsan-si (eastern part of Korea) with weak BS. It could provide crucial information for origin of Korean endemic, *G. verum* var. *asiaticum* f. *pusillum*. It needs additional study to investigate the origin and in these evolutionary relationship among these taxa. The four taxa of *G. dahuricum* group from Korea-Japan; *G. dahuricum* var. *lasiocarpum*, *G. kikumugura*, *G. tokyoense,* and *G. pseudoasprellum*, are have been included into sect. *Trachygalium* (Ehrendorfer *et al*., 2005). There is no study of phylogenetic using molecular makers before. The four taxa of *G. dahuricum* group occur in East Asia, and have serious identification problems and taxon delimitation due to severe variations in the morphology of leaves, seed hairs and flower and inflorescences (Chen and Enrendorfer, 2011). We confirmed the phylogenetic relationship among these taxa, for the first time. *G. kikumugura* and *G. tokyoense* were included in Group C. *G. kikumugura* having whorls of four leaf-like organs and fruit with generally hooked hairs were closely related to *G. tokyoense,* morphologically (Yamazaki, 1993). Lee (1995) reported the distribution of *G. kikumugura* in Korea but we could not find the distibution during the this study although the species is widely distributed in Japan. We also could not confirm *G. kikumugura* specimens collected from Korea at Korean and Japan herbria. Therefore we assumed that the distribution report of this taxa by Lee (1995) was based on misclassification. *G. pseudoasprellum* was treated as synonyms of *G. dahuricum* by Ehrendorfer *et al.* (2005), but in our results did not support his opinion. *G. pseudoasprellum* is similar to *G. dahuricum* var. *lasiocarpum,* morphologically but it can be distinguished from *G. dahuricum* based on leaf shapes, which whorl of 6 elliptic or lanceolate leaves. *G. tokyoense* has glabrous fruit and white flower compare with *G. dahuricum* var. *lasiocarpum*. Previous studies based on morphology (Yamazaki, 1993; Pobedimova *et al*., 2000; Chen and Ehrendorfer, 2011) were argument for classification of *G. tokyoense*. We confirmed that the *G. tokyoense* and *G. dahuricum* var. *lasiocarpum* were polyphyletic. Also our result is supported that previous classification that *G. tokyoense* be regarded as a species. *G. kamchaticum* Steller *ex* Schultes & J. H. Schultes and *G. kamchaticum* var. *yakusimense* (Masamune) Yamazakiwere place to clade□ with polytomy at MP tree with weak PP value. *G .kamchaticum* is distributed in an alpine meadow of worldwide with centers of the diversity in eastern Asia and eastern North America (Ehrendorfer *et al*., 2005). *G. kamchaticum* var. *yakusimense* is smaller leave and tall than *G. kamchaticum*. This species is erect, with round leaves, one vein, whorls of four leaf-like organs, 4-parted white, and a fruit with generally hooked hairs.

We confirmed that *G. kinuta* is closer to *G. boreale*. Two taxa usually occur in northern part of Korean peninsula, especially in the mountain forests in lower elevation. The somatic chromosome number of *G. kinuta* and *G. boreale* were 4X ($2n$=44) and/or 2X ($2n$=11), respectively (Jeong and Pak, 2009). *G. kinuta* is erect, four leaf-like organs, three veins, branched panicles of inflorescences, and white flowers. *G. kinuta* and *G. boreale* are generally very similar in morphology and can be distinguished by the characters of leaf-shape.

Table 2. Primers used for amplification of cpDNA regions in this study.

Region	Primer	Sequence (5'-3')	Annealing temperature (°C)	References
*trnC-ycf*6	trnC^{GCA}F	CCAGTTCRAATCYGGGTG	52	Demesure *et al.* (1995)
	ycf6R	GCCCAAGCRAGACTTACTATATCCAT		Demesure *et al.* (1995)
*trn*F-*ndh*J	ndhJ	ATGCCYGAAAGTTGGATAGG	57	Shaw *et al.* (2007)
	TabE	GGTTCAAGTCCCTCTATCCC		Taberlet *et al.* (1991)
*Trn*L intron	c	CGAAATCGGTAGACGCTACG	55	Taberlet *et al.* (1991)
	d	GGGGATAGAGGGACTTGAAC		Taberlet *et al.* (1991)
*rpo*B-*trn*C	rpoBb	CGGATATTAATAKMTACATACG	55	Soza and Olmstead (2010)
	rpoBd	GTTGGGGTTTACATATACT		Soza and Olmstead (2010)

The *G. trachygalium* group consisted of four species; *G. trachygalium*, *G. pogonanthum*, *G. gracilens*, which occur in both Korea and Japan, and *G. koreanum* endemic to Korea. Although, the four taxa placed into Group A, our data did not provide insights into the specific phylogenetic relationships among *G. trachygalium* group species. These taxa are characterized by whorls of four leaf-like organs, cymose inflorescences with several terminal flowers, 4 parted rotate flowers and tuberculate fruit. The identification and delimitation of these species are usually difficult because they are very similar in morphology. The four species are distinguished by the differences in leaf size, shape, and fruit hairs (Jeong and Pak, 2012). These taxa usually occur in the near or same population, and share a common habitat. The somatic chromosome number of these species are 2X ($2n$=22) and/or 4X ($2n$=44) (Jeong and Pak, 2009). This inconsistencies phylogeny can be explained the speciation processes of the *G. trachygalium*group. But it is yet to be determined whether incomplete lineage sorting of ancestral polymorphisms in the population, or chloroplast capture by hybridization and introgression. It needs additional study to understand origin and clear relationship among these taxa. *G. trachyspermum*, *G. pogonanthum* and *G. gracilens* previously been placed into a sect. *Cymogalia* by Yamazaki (1993) but our data showed that these four taxa including *G. koreanum*, are more closely related to members of sect. *Platygalium* (Table 1). We suggest that the four taxa have to be transferred to sect. *Platygalium* based on molecular and morphological data.

Acknowledgement

This research was supported by Basic Science Research Program through the National Research Foundation of Korea (NRF) funded by the Ministry of Education (2016R1A 6A1A05011910).

References

Chen, T. and Ehrendorfer, F. 2011. Rubia.Vol 19. In: Wu ZY, Raven PH, Hong DY(Eds). Flora of china. Beijing: Science Press; St. Louis: Missouri Botanical Garden Press. pp. 104-141.

Demesure, B., Sodzi, N., Petit, R.J. 1995. A set of universal primers for amplification of polymorphic non-coding regions of mitochondrial and chloroplast DNA in plants. Molec. Ecol. **4**: 129-131.

Dolye, J.J. and Dolye, J.L.1987. A rapid DNA isolation procedure for small quantities of fresh leaf tissue.Phytochem. Bul. Bot. Soc. Amer. **19**:11-15.

Ehrendorfer, F., Manen, J.-F.and Natali, A. 1994. Cp DNA intergene sequences corroborate restriction site data for reconstructing Rubiaceae phylogeny. Pl. Syst. Evol. **190**: 195-211.

Ehrendorfer, F., Schönbeck-Temesy, E., Puff, C. and Rechinger, W. 2005. Rubiaceae.eds. K. H. Rechinger, Flora Iranica. no. 176. Verlag des Naturhistorischen Museums Wien, Vienna, Austria.

Ehrendorfer, F., Vladimirov, V. and Barfuss, M.H.J. 2014. Paraphyly and polyphyly in the worldwide tribe Rubieae (Rubiaceae): Challenges for genetic delimitation. Ann. Missouri Bot. Gard. **100**: 79-88.

Felsensteijn, J. 1985. Confidence limits on phylogenies: an approach using the bootstrap. Evolution. **39**: 783-791.

Govaerts, R. 2006. World checklist of selected plant families. In: F. A. Bisby, Y.R. Roskov, M. A. Ruggiero, T. M. Orrell, L. E. Paglinawan, P. W. Brewer, N.Bailly, and J. van Hertum, (eds.), Species 2000 & ITIS catalogue of life: www.catalogueoflife.org/annualchecklist/ 2007/. Species 2000, Reading, U. K.

Jeong K.S. and Pak, J. H. 2009. A cytotaxonomic study of *Galium* (Rubiaceae) in Korea. Korean J. Pl. Tax. **39**: 42-47.

Jeong K.S. and Pak, J. H. 2012. The morphological study of *Galium* L. (Rubiaceae) in Korea. Korean J. Pl. Tax. **42**: 1-12.

Lee, T.B. 1979. Illustrated Flora of Korea. Hyangmunsa, Seoul (in Korean).

Lee, W.T. 1995. Lineamenta Florae Koreae. Academy Press, Seoul (in Korean).

Lee, Y.N. 2004. Flora of Korea. Kyohaksa, Seoul (in Korean).

Manen, J.-F., Natali, A. and Ehrendorfer, F. 1994. Phylogeny of Rubiaceae-Rubieae inferred from the sequence of a cpDNA intergene region.Pl. Syst. Evol.**190**: 195-211.

Manen, J.-F.and Natali, A. 1995. Comparison of the evolution of ribulose-1,5-biphosphate carboxylase (*rbcL*)and *atb-rbcL* noncoding spacer sequences in a recent plant group, the tribe Rubieae (Rubiaceae). J. Molec. Evol. **41**: 920-927.

Natali, A., Manen, J.-F.andEhrendorfer, F. 1995. Phylogeny of the Rubiaceae-Rubioideae, in particular the tribe Rubieae: Evidence from a non-coding chloroplast DNA sequence. Ann. Missouri Bot. Gard. **82**: 428-439.

Natali, A., Manen, J.-F., Kiehn, M. and Ehrendorfer, F. 1996. Tribal, generic, and specific relationships in the Rubioideae-Rubieae (Rubiaceae) based on sequence data of a cpDNA intergene region. Opera Bot. Belg. **7**: 193–203.

Nylander, J.A.A. 2004. Mrmodeltest 2.3. Program distributed by the author. Evolutionary biology Centre, Uppsala University.

Pobedimova, E.G. 2000. *Galium* L. in: B. K. Schischkin (ed.), Flora of the U.S.S.R, vol. 23. Bishen Singh Mahendra Pal Singh, Dehra Dun. India, and Koeltz Scientific Books, Koenigstein, Germany. pp. 345–459.

Robbrecht, E. and Manen, J.F. 2006. The major evolutionary lineages of the coffee family (Rubiaceae, angiosperms). Combined analysis (nDNA and cpDNA) to infer the position of Coptosapelta and Luculia, and supertree construction based on *rbcL*, *rps*16, *trn*L-*trn*F and *atp*B-*rbc*L data. A new classification in two subfamilies, Cinchonoideae and Rubioideae. Syst. Geogr. Plants. **76**: 85-146.

Ronquist, F. and Huelsenbeck, J.P. 2003. Mrbayes 3: Bayesian phylogenetic inference under mixed models. Bioinfomatics. **19**: 1572-1574.

Shaw, J., E. B. Lickey, J. T. Beck, S. B. Farmer, W. Liu, J. Miller, K. C. Siripun, C. T. Winder, E. E. Schilling, and R. L. Small. 2005. The tortoise and the hare II: Relative utility of 21 noncoding chloroplast DNA sequences for phylogenetic analysis. Am. J. Bot. **92**: 142-166.

Shaw, J., Lickey, E.B., Schililling, E.E. and Small, R.L. 2007. Comparison of whole chloroplast genome sequences to choose noncoding regions for phylogenetic studies in angiosperms; the tortoise and the hare. Am. J. Bot. **94**: 275-288.

Soza, V.L. and Olmstead, R.G. 2010. Molecular systematics of tribe Rubieae (Rubiaceae): Evolution of major clades, development of leaf-like whorls, and biogeography. Taxon **59**: 755-771.

Swofford, D.L. 2003. PAUP: Phylogenetic analysis using parsimony (and other method). Ver.4.0b10. Sinauer Associates, Sunderlad, Massachusetts.

Taberlet, P., Gielly, L. Pautou, G. and Bouvet, J.1991. Universal primer for amplification of three non-coding regions of chloroplast DNA.Pl.Molec. Biol. **17**: 1105-1109.

Thompson, J.D., Gibson, T.J., Plewniak, F., Jeanmougin, F. and Higins, D.G. 1997. The CLUSTAL X windows interface: flexible strategies for multiple sequence alignment aided by quality analysis tools. Nucleic Acids Res. 22: 4676-4882.

Yamazaki, T. 1993. Angiospermae. Vol□a. *In:* Iwatsuki K., David E.B., Ohba H. (Eds.) Flora of Japan. Tokyo: Kodansa. pp. 233-240.

18

MORPHOLOGICAL VARIABILITY OF EVERGREEN OAKS (*QUERCUS*) IN TURKEY

YILMAZ AYKUT*, USLU EMEL[1] AND BABAÇ M. TEKIN[1]

Department of Molecular Biology and Genetics, Faculty of Science and Arts, Uşak University, 64200 Uşak, Turkey

Keywords: *Ilex;* Morphometric, UPGMA; Turkey.

Abstract

The genus *Quercus* L. has a problematic taxonomy because of widespread hybridization among them. Evergreen *Quercus* contain three species in section *Ilex* Loudon namely, *Q. ilex* L., *Q. coccifera* L. and *Q. aucheri* Jaub. *et* Spach in Turkey. Here, two species, *Q. coccifera* and *Q. aucheri* are usually confused with each other. However, *Q. coccifera* and *Q. calliprinos* are accepted as different species but this subject is still controversial. Morphometric leaf and fruit variations of *Q. ilex*, *Q. coccifera* and *Q. aucheri* in 26 populations were measured for 25 characters. Variations within and among populations of species were detected by cluster analysis and principal component analysis. This study shows that populations of *Q. coccifera* from the south region of Turkey form a second group within *Q. coccifera*. Secondly, *Q. coccifera* show more similarity to *Q. aucheri* than *Q. ilex*, and finally there are two groups within *Q. coccifera*, which may be evaluated as *Q. coccifera* and *Q. calliprinos*.

Introduction

The genus *Quercus* L., popularly known as oaks shows highest morphological variations among species and populations (Hokanson *et al.*, 1993; Kremer and Petit, 1993), especially its leaf characters are the most valuable in the classification and delimitation of species (Borazan and Babaç, 2003). The major reason for the phenotypic diversification of oaks is the high frequency of hybridization among species (Borazan andBabaç, 2003; Jensen, 1995).

Leaves are good indicators of putative hybridization and oaks can be easily identified by their leaves. The leaves of hybrid species have typically asymmetric shapes and are irregular (Jensen, 1995). Because of common interspecific hybridization in the genus *Quercus*, individuals that exhibit intermediate morphological characters can be seen widely. Sometimes, it is not possible to identify oak species due to high morphological variation. In this case, acorns are secondary important materials in oaks for classification and determination of hybridization (Jensen, 1988).

In Turkey three evergreen *Quercus* species exist, viz. *Q. aucheri*, *Q. coccifera* and *Q. calliprinos*. *Q. aucheri* is not very widely distributed, it only exists in the south western Anatolia region of Turkey and in the Greek islands like Rhodos. *Q. coccifera* is confused with *Q. calliprinos* Webb. (Toumı and Lumaret, 2001) because of small or medium shrub formation and acorn shape and evergreen nature. These two species may be evaluated within *Q. coccifera* as *Q. coccifera* subsp. *coccifera* and *Q. coccifera* subsp. *calliprinos* (Webb) Holmboe as two different species. This subject is still controversial (Salvatore and Paola, 1976; Toumı and Lumaret, 2001). *Q. ilex* the last member of *Ilex* section has two morphological types. These are the rotundifolia type containing small and round leaves and the ilex type containing big pointed leaves. The

---

*Corresponding author: Email:aykutyilmaza@gmail.com
[1]Department of Biology, Faculty of Science and Arts, Abant Izzet Baysal University, Bolu, Turkey.
DOI: http://dx.doi.org/10.3329/bjpt.v24i1.33004

rotundifolia morphotype exists in North Africa and the interior region of Spain (Tutın *et al.*, 1964). The ilex morphotype exists along the Atlantic coast of France. These two morphotypes are two different species (Tutın *et al.*, 1964) or two subspecies (Saenz De Rıvas, 1967) or only two varieties (Maıre, 1961). Additionally, the presence of intermediate forms for these two morphotypes is reported in the Mediterranean region of France and in the north and east coasts of Spain. The aims of the present study were firstly to examine the morphological relationships among the populations sampled and their potential hybrids in Turkey, secondly to designate the status of confused two species, *Q. coccifera* and *Q. calliprinos*, and finally to evaluate and compare results provided from the leaf and acorn character.

Materials and Methods

The populations sampled are located in the 17 provinces that include the regions of North West, West, South and South West of Turkey. A total of 26 populations belonging to three species of *Ilex* section, namely *Quercus coccifera*, *Q. ilex* and *Q. aucheri* were collected. While 16 populations were designated to reveal variations within *Q. coccifera* (Table 1 and Fig. 1). *Q. aucheri* and *Q. ilex* were sampled in 5 populations owing to their distributions in a restricted region.

Table 1. Study populations with population number, location, coordinates and altitude (C = *Q. coccifera*, A = *Q. aucheri*, I = *Q. ilex*) *Q. coccifera* (C), *Q. aucheri* (A) and *Q. ilex* (I).

Q. coccifera (C), Q. aucheri (A) and Q. ilex (I)			Altitude (m)
Pop. No.	Location	Coordinates	
		N	E	
C_1	İzmir-Balıkesir border area, Altınova barrage road	$39^0$12.903	$026^0$49.302	70
C_2	İzmir-between Dikili-Çandarlı, 20 km. to Çandarlı	$39^0$01.253	$026^0$55.505	40
C_3	Manisa-between Kırkağaç-Akhisar, 1-2 km. after Çandarlı	$39^0$05.800	$027^0$40.257	190
C_4	Çanakkale-Ezine-Bozcaada pier	$39^0$47.950	$026^0$12.115	50
C_5	Gökçeada-between Gökçeada-Dereköy	$40^0$09.689	$025^0$49.586	60
C_6	Mersin-5-10 km. after Seratvul	$36^0$50.997	$033^0$18.402	1400
C_7	Karaman-between Mut-Ermenek, 45 km. before Ermenek	$36^0$37.276	$032^0$55.182	1300
C_8	Antalya-between Korkuteli-Bucak, 25 km. before Bucak	$37^0$15.582	$030^0$19.362	920
C_9	Aydın-Eski Çine, Ovacık village	$37^0$32.889	$028^0$05.310	300
C_{10}	Aydın-Söke, between Bağarası-Akçakaya village	$37°40.350$	$027°31.347$	40
C_{11}	Muğla-between Muğla-Kale, 59 km. before Kale	$37^0$08.142	$028^0$32.157	800
C_{12}	Denizli- between Kale-Tavas, 1-2 km. before Tavas	$37°33.069$	$029°03.150$	940
C_{13}	Uşak-between Sivaslı-Uşak, 12 km. after Sivaslı	$38°34.259$	$029°36.303$	825
C_{14}	Gaziantep- between Yavuzeli-Araban	$37^0$22.975	$037^0$33.292	740
C_{15}	Kahramanmaraş- between k.maraş- göksun	$37^0$43.514	$036^0$40.038	1075
C_{16}	Hatay-between Kırıkhan-Hassa	$36^0$36.554	$036^0$23.591	350
A_1	Antalya-between Kemer-Kumluca	$36^0$25.429	$030^0$25.447	530
A_2	Aydın-Çine,Across from the cemetery Kuruköy	$37^0$33.558	$028^0$04.047	180
A_3	Aydın-Priene-Söke	$37^0$44.967	$029^0$16.369	90
A_4	İzmir-Selçuk-Zeytinköy	$37^0$59.569	$027^0$17.226	65
A_5	Muğla-between Milas-Bodrum, Dörttepe village	$37^0$11.242	$027^0$37.142	8
I_1	Zonguldak-Alaplı, Sabırlı village	$41^0$08.901	$031^0$23.147	180
I_2	Zonguldak-between Alaplı-Düzce	$41°08.443$	$031°20.596$	4
I_3	Düzce- between Yığılca-Alaplı	$41^0$09.136	$031^0$23.627	60
I_4	İstanbul-between Anatolian Fortrees-Kavacık	$41^0$04.220	$029^0$05.085	65
I_5	Gökçeada-between Gökçeada-Dereköy	$40^0$09.689	$025^0$49.586	60

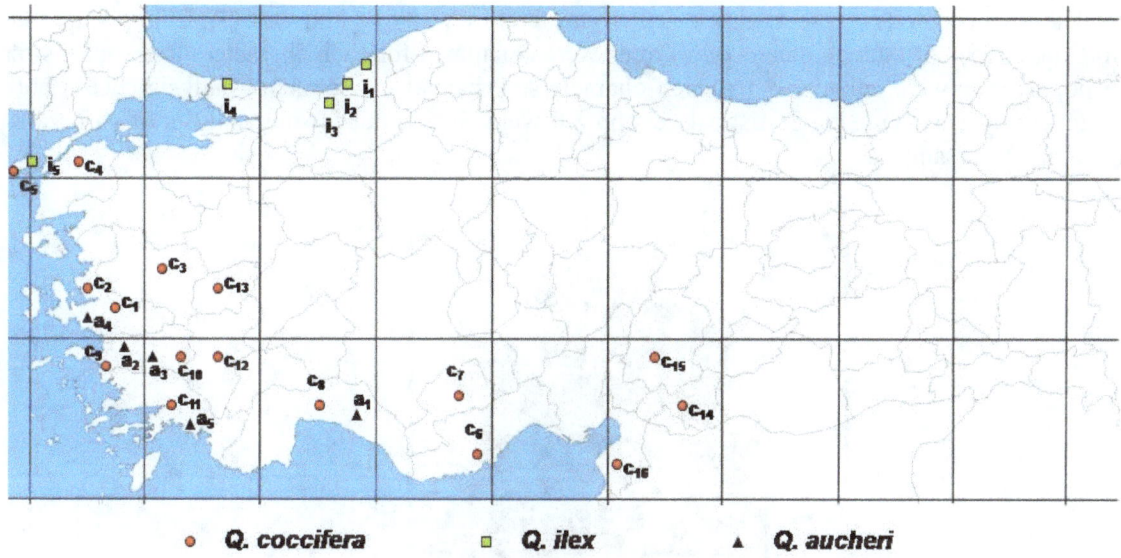

Fig. 1. Distribution of studied populations of *Q. coccifera, Q. ilex* and *Q. aucheri* in Turkey.

Leaf and fruit samples for identification and statistical analyses of each population were collected from 260 trees. In total, 2600 leaf and fruit materials were measured. All leaf samples were collected at the same height and location after leaf growth had stopped to avoid seasonal and positional variations as reported by Blue and Jensen (1988). Fruit samples were selected from mature acorns and cupules were also used as fruit characters.

Ten characters of leaves (Table 2 and Fig. 2) ad 15 characters from fruits (Table 3 and Fig. 3) were used. Most of the leaf characters were adopted from different sources (Bruschı *et al.*, 2000; Kremer *et al.*, 2002; Borazan and Babaç, 2003; Bruschı *et al.*, 2003; Gonzalez-Rodrıguez *et al.*, 2004; Ponton *et al.*, 2004; Boratynskı *et al.*, 2008).

Table 2. The leaf characters used in the morphological analysis.

LL	:	Lamina length
PL	:	Petiole length
MWL	:	Maximal width of lamina
MW	:	Middle width of lamina
DTW	:	The distance between the widest point and the leaf tip
DBW	:	The distance between the widest point and the leaf base
TLL	:	Total leaf length (LL+LP)
P%	:	Petiole length (PL) x 100/total leaf length (TLL)
MW%	:	Middle width of lamina (MW) x 100/total leaf length (TLL)
MWL%	:	Maximal width of lamina (MWL) x 100/total leaf length (TLL)

Table 3. The fruit characters used in the morphological analysis.

Nut characters	Cupsule characters	Stalk characters	Index characters
1. Nut length	4. Cupule outer diameter	9. Stalk length	11. Cupule length/Nut length
2. Nut diameter	5. Cupulethickness	10. Stalk thickness	12. Cupule depth/Cupule length
3. Nut scar diameter	6. Cupule scale length (maximum)		13. Nut diameter/Nut length
	7. Cupule depth		14. Cupule thickness/Cupule outer diameter
	8. Cupule length		15. Acorn mass

The fruit characters were also selected from literature (Nikoliç and Orloviç, 2002; Tılkı and Alptekin, 2005). Arithmetic means of all trees were calculated for each character. Then, means of the populations were calculated for each characters. Principal Component Analysis (PCA) and Cluster Analysis (CA) using Statistical version 8.0 were carried out for the analysis of variations in leaf and fruit samples.

Fig. 2. Morphological leaf characters

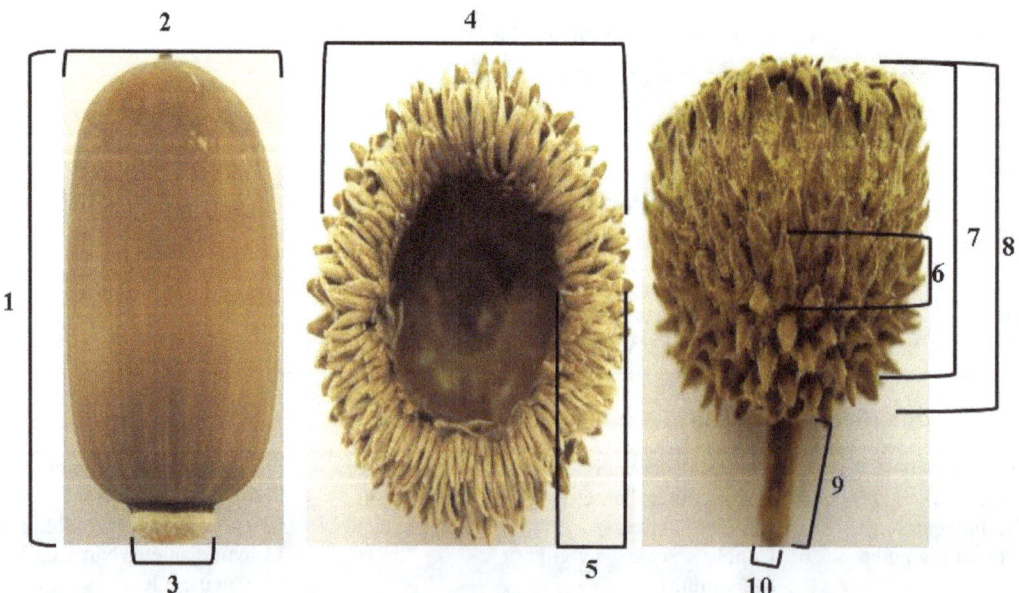

Fig. 3. Morphological fruit characters with character number.

Results

The UPGMA cluster analysis performed on the populations for leaf characters recognized two main groups one contains all populations of *Q. ilex* and another group consisted of 21 populations belonging to *Q. coccifera* and *Q. aucheri*. (Fig. 4) When the first group having the populations of *Q. ilex* is evaluated, it can be stated that populations of *Q. ilex* tend to form more morphologically discrete group than *Q. coccifera* and *Q. aucheri* populations and geographically close populations show more similarity like İ₁ and İ₂ populations (Fig. 4) Differences in geographical distribution are effective on species diversity. The biggest difference in the populations of *Q. ilex* is observed in İ₅ population and the locality of this population is an island in Aegean sea.

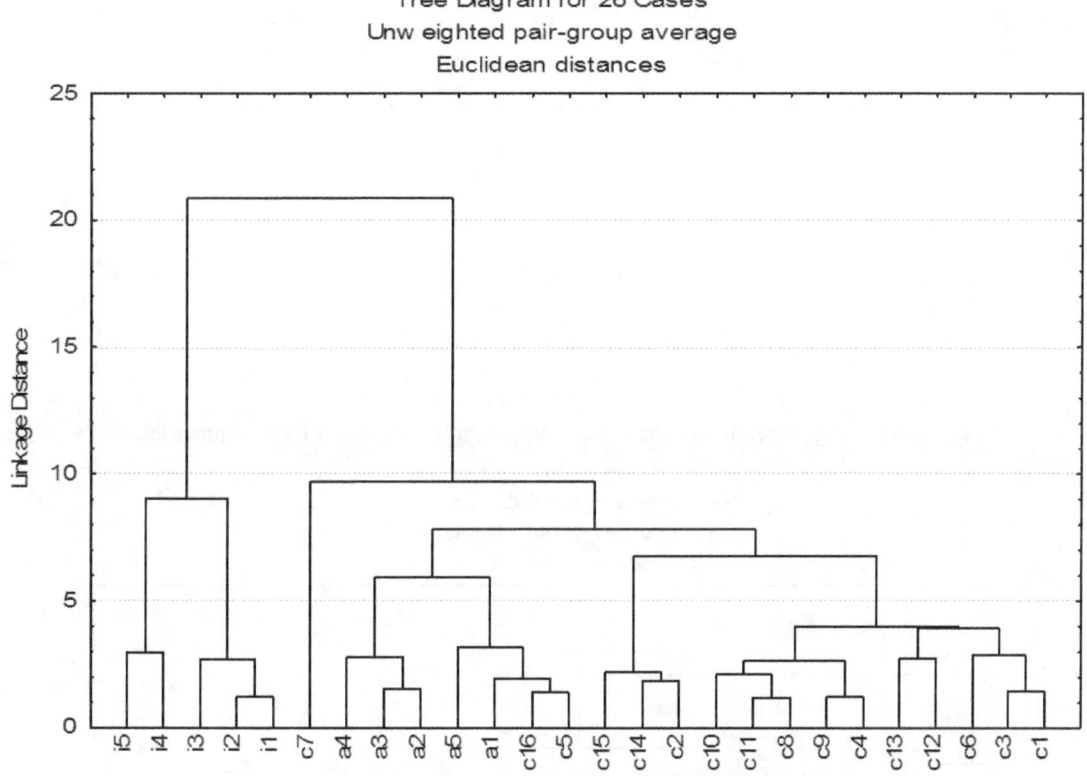

Fig. 4. Phenogram resulting from Cluster Analysis with UPGMA for the leaf materials.

The largest variation in second main group comprising of 21 populations is observed in C_7 population of *Q. coccifera* that occurs at high altitude (Table 1 and Fig. 4). Except C_7 population, the second main group is divided into two sub-groups, one consists of all populations of *Q. aucheri* and only C_{16} and C_5 populations of *Q. coccifera*, while other sub-group consists of complete populations of *Q. coccifera* (Fig. 4). The highest variation within this sub-group are observed in C_{14}, C_{15} and C_2 populations. Populations of *Q. aucheri* show the differences from populations of *Q. coccifera* within the second main group but this difference is not clear as in *Q. ilex*.

PCA results (Fig. 5) show the high similarity with CA (Fig. 4) results. PCA analyses clearly separate the *Q. ilex* populations from others. Similarly, two main groups are observed from PCA analysis. While one of these groups consists of populations belonging to *Q. ilex*, other two species are evaluated in the second group. C_{14}, C_{15} and C_{16} populations show the most differences within *Q. coccifera*. Results revealed that fruit characters, *Q. ilex*is is separated from the other two species (Figs 6 & 7). CA results show clearly the presence of two main groups. The first main group consists of populations of *Q. ilex* and the second main group consists of the populations of *Q. coccifera* and *Q. aucheri*. This result shows the high similarity with the results of leaves.

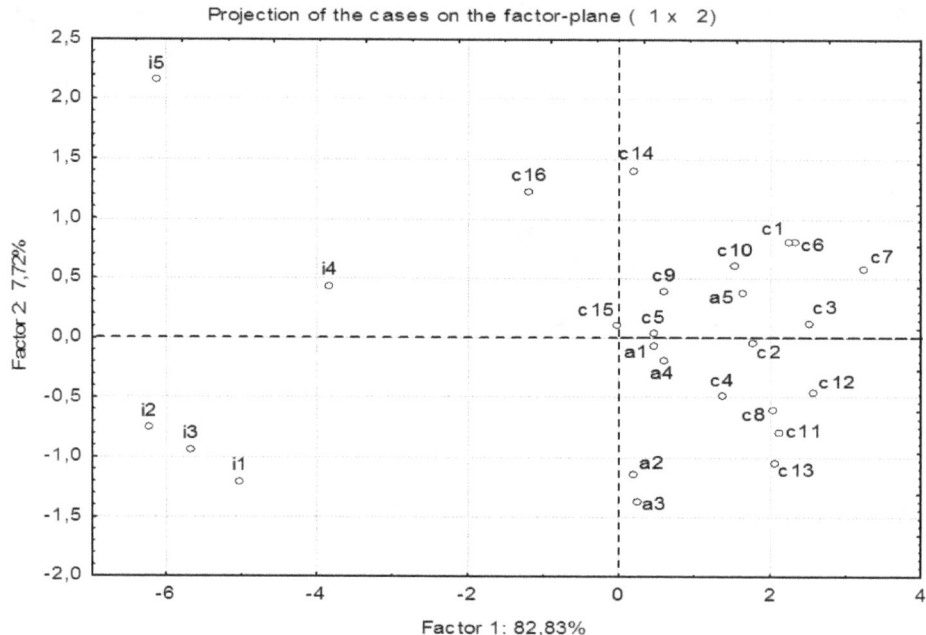

Fig. 5. Resulting projection of Principal Component Analysis for the leaf materials.

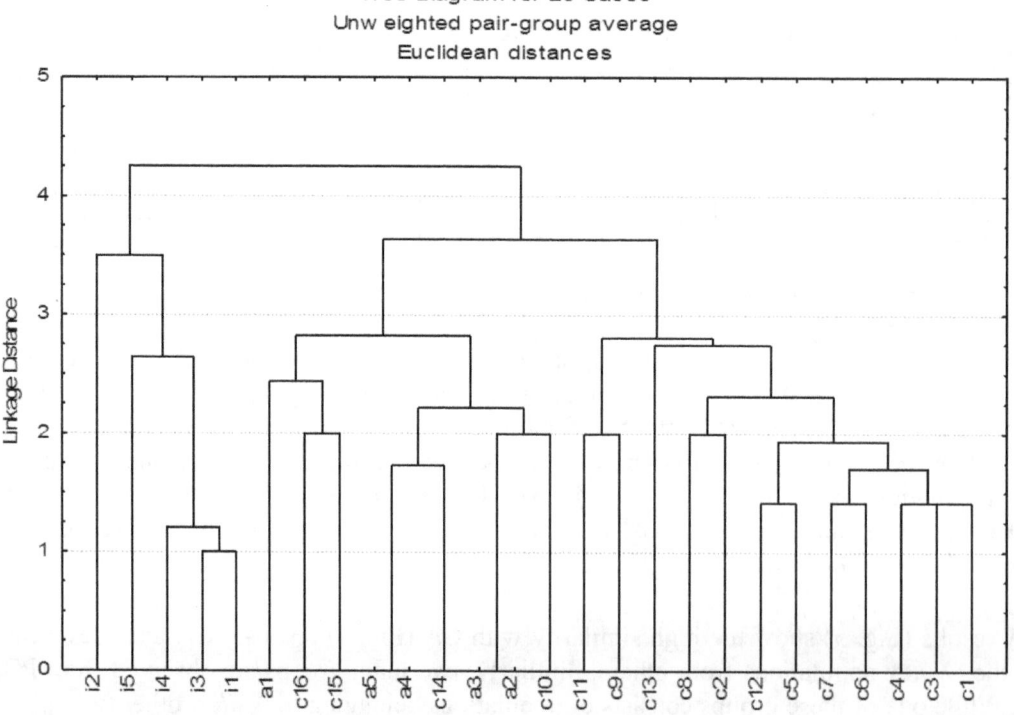

Fig. 6. Phenogram resulting from Cluster Analysis with UPGMA for the fruit materials.

Populations C_{10}, C_{14}, C_{15} and C_{16} form a discrete group with *Q. aucheri* in the CA graph (Fig. 6). However, remaining populations of *Q. coccifera* form other group (Fig. 6). Similar results showing differences among the species are observed in PCA (Fig. 7).

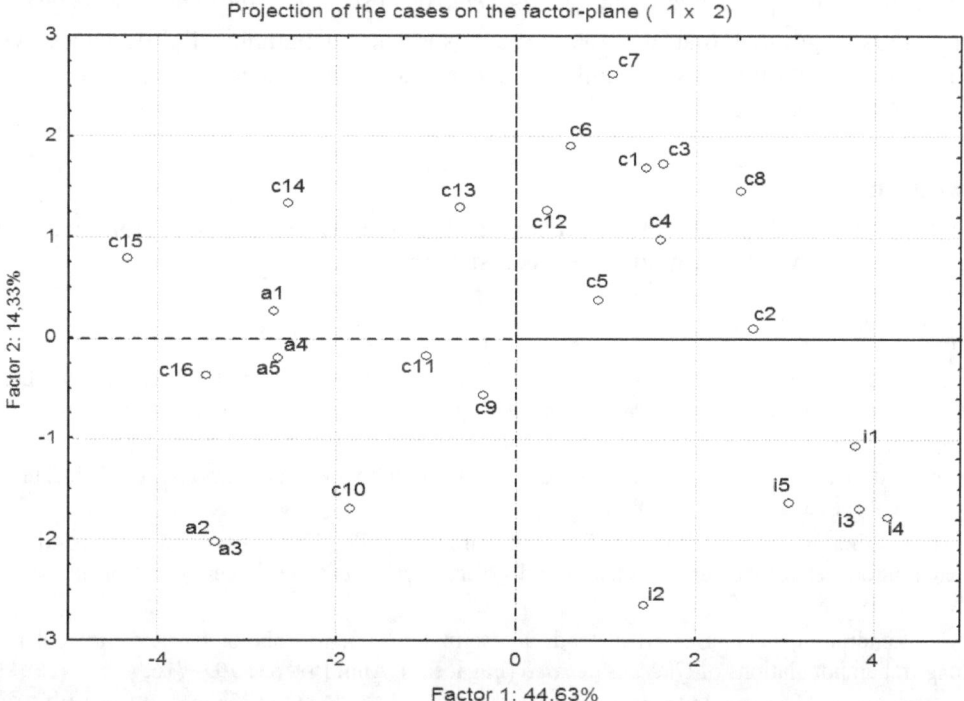

Fig. 7. Resulting projection of Principal Component Analysis for the fruit materials.

Discussion

The present study revealed that PCA and CA analyses could be used to solve taxonomic problems and to understand the relations among three species belonging to *Ilex* section of the genus *Quercus*. Leaf materials were generally used for the comparison of the oaks (Borazan and Babaç, 2003; Bruschı *et al.*, 2003; Ponton *et al.*, 2004; Gonzalez-Rodrıguez and Oyama, 2005; Franjıc *et al.*, 2006; Boratynski *et al.*, 2008) but here, for the first time the fruit materials together with the leaf materials were examined in detail.

The results of morphometric studies provided the satisfactory findings for phenetic groupings of taxa in *Ilex* section. The most significant differences were found on the *Q. ilex* populations. This species was separated from the remaining species on the basis both leaf and fruit. However, *Q. coccifera* populations were grouped next to *Q. aucheri* populations. On the other hand, the results of *Q. coccifera* and *Q. aucheri* are not clearly separated from each other. Especially, due to similar leaf and fruit characters in both taxa, they showed introgression with each other in both CA and PCA plots (Figs 4-7). However, these results draw attention to the presence of a group away from *Q. coccifera*. The first group consists of the populations sampled from North West, West and South West regions of Turkey. The populations sampled from the south region of Turkey such as C_{14}, C_{15} and C_{16} was included into the second group showing the similarity to the populations of *Q. aucheri*. Result from the leaf and fruit studies supported these groupings. Similar results are also observed by Salvatore and Paola (1976), Toumı (1995), ToumıandLumaret (2001)and Yılmaz *et al.*(2013).

Geographically separation of *Q. coccifera* suggests that there are variation within this species. The restricted group of *Q.coccifera* located only in the south region of Turkey is geographically closer to Syria, Israel and Palestine. In Palestine there are two subspecies *Q. calliprinos* Webb. viz. *Q. calliprinos* sub sp. *coccifera* and *Q. calliprinos* sub sp. *calliprinos* (Zohary, 1966).

Our results suggested that the two groups showing geographical differences within *Q. coccifera* may be quite possibly strengthen the existence of two species as *Q. coccifera* and *Q. calliprinos* (Yılmaz *et al.*, 2013).

Acknowledgements

The authors would like to thank Abant İzzet Baysal University Directorate of Scientific Research Projects (BAP) for providing financial support.

References

Boratynskı, A.,Marcysıak, K., Lewandowska, A., Jasınska, A., Iszkulo, G. And Burczyk, J. 2008. Differences in leaf morphology between *Quercus petraea* and *Q. robur* adult and young individuals. Silva Fenn. **42:** 115–124.

Borazan, A. and Babaç, M.T. 2003. Morphometric leaf variation in oaks (*Quercus*) of Bolu,Turkey. Ann. Bot. Fenn. **40:** 233–242.

Bruschı, P.,Vendramın, G.G., Busottı, F. And Grossonı, P. 2000. Morphological and molecular differentiation between *Quercus petraea* and *Quercus pubescens* (Fagaceae) in Northern and Central Italy. Ann. Bot. **85:** 325–333.

Bruschı, P., Vendramın, G.G., Busottı, F. and Grossonı, P. 2003. Morphological and molecular diversity among Italian populations of *Quercus petraea* (Fagaceae). Ann. Bot. **91:** 707–716.

Gonzalez-Rodrıguez, A., Arıas, D.M,,Valencıa, S. and Oyama, K. 2004. Morphological and RAPD analysis of hybridization between *Quercus affinis* and *Q. laurina* (Fagaceae), two Mexican red oaks. Am. J. Bot. **91:** 401–409.

Gonzalez-Rodrıguez, A. and Oyama, K. 2005. Leaf morphometric variation in *Quercus affinis* and *Q. laurina* (Fagaceae), two hybridizing Mexican red oaks. Bot. J. Linn. Soc .**147:** 427–435.

Hokanson, S.C., Isebrands, J.G., Jensen, R.J. and Hancock, J.F. 1993. Isozyme variation in oaks of the Apostle Islands in Wisconsin: Genetic structure and levels of inbreeding in *Quercus rubra* and *Quercus ellipsoidalis* (Fagaceae). Am. J. Bot. **80:** 1349–1357.

Jensen, R.J. 1988. Assesing patterns of morphological variation of *Quercus* spp. in mixed-oak communities. Am. Midl. Nat. **120:** 120–135.

Jensen, R.J. 1995. Using leaf shape to identify taxa in a mixed-oak community in land between the lakes, Kentucky. Proc. Sixth Symposium Nat. Hist. lower Tennessee and Cumberland River Valleys 177-188. Center Field Biol., Austin Peay State Univ., Clarksville.

Kremer, A., Dupouey, J.L., Deans, J.D., Cottrell, J., Csaıkl, U., Fınkeldey, R., Espınel, S., Jensen, J., Kleınschmıt, J., Van Dam, B., Ducousso, A., Forrest, I., De Heredıa, U.L., Lowe, A.J., Tutkova, M., Munro, R.C., Steınhoff, S. and Badeau, V. 2002. Leaf morphological differentiation between *Quercus robur* and *Quercus petraea* in stable across western European mixed oak stands. Ann. Forest Sci. **59:** 777–787.

Kremer, A. and Petıt, R.J. 1993. Gene diversity in natural populations of oak species. Ann. Forest Sci. **50:** 186–202.

Maıre, R. 1961. Flore de l'Afrique du Nord, Vol. 7. Ed Lechevallier, Paris.

Nıkolıc, N.P. and Orlovıc, S.S. 2002. Genotypic variability of morphological characteristics of English oak (*Quercus robur* L.) acorn. Proceeding for Natural Sciences **102:** 53–58.

Ponton, S., Dupouey, J.L. and Dreyer, E. 2004. Leaf morphology as species indicator in seedlings of *Quercus robur* L. and *Q. petraea* (Matt.) Liebl.: modulation by irradiance and growth flush. Ann. Forest Sci .**61**: 73–80.

Saenz De Rivas, C. 1967. Estudios sobre *Quercus ilex* L., *Quercus rotundifolia* Lamk. Anales Del Instituto Botanico A. J. Cavanilles **25**: 243–262.

Salvatore, G. and Paola, G. 1976. *"Quercus calliprinos"* Webb e *"Quercus coccifera"*L.:Ricerche sull'anatomia fogliare e valutazioni tassonomiche e corologiche. Giornale Botanico Italliano **110**: 89–115.

Tılkı, F. and Alptekın, C.U. 2005. Variation in acorn characteristics in three provenances of *Quercus aucheri* Jaub. et Spach and provenance, temperature and storage effects on acorn germination. Seed Sci. Technol. **33**: 441–447.

Toumı, L. 1995. Etude de la structure genetique et introgressions eventuelles chez les chenes sclerophylles mediterraneens a l'aide de marquers alloenzymatiques. Ph. D. Thesis, University of Aix-Marsielle III, Marsielle.

Toumı, L. and Lumaret, R. 2001. Allozyme characterization of four Mediteranean evergreen oak species. Biochem. Syst. Ecol. **29**: 799–817.

Tutın, T.G., Heywood, V.H., Burges, N.A., Moore, D.M., Valentıne, D.H., Walters, S.M. and Webb, D.A. 1964. Flora Europaea. Cambridge University Press, London.

Yılmaz, A., Uslu, E. and Babaç M.T. 2013. Molecular diversity among Turkish oaks (*Quercus*) using random amplified polymorphic DNA (RAPD) analyses. Afr. J. Biotechnol.**12**: 6358-6365.

Zohary, M. 1966. Flora Palaestina. Jerusalem Academic Press. Israel.

19

MUSA × *PARAHAEKKINENII* (MUSACEAE): A NEW ARTIFICIAL INTERSPECIFIC HYBRID FROM KERALA, INDIA

Komban Parameswaran Smisha and Mamiyil Sabu[1]

Angiosperm Taxonomy and Floristics Division, Department of Botany, University of Calicut, Kerala 673 635, India

Keywords: *Musa coccinea*; *Musa haekkinenii*; Wild parents; Manual crosses; New artificial hybrid.

Abstract

Musa× parahaekkinenii (Musaceae), a new manually crossed interspecific hybrid of two wild parent plants *Musa coccinea* Andrews (female) and *Musa haekkinenii* N.S. Lý & Haev. (male), is described and illustrated. A comparison of characters with its parents and a key to the new hybrid *M. × parahaekkinenii* are provided.

Introduction

The Musaceae (Commelinids: Zingiberales) is a tropical family comprising three genera, namely *Ensete* Horaninow, *Musa* L. and *Musella* (Franchet) Wu. *Musa* L. is the largest genus of the family (ca. 65 species) distributed in tropical Asia from southern India to eastern Himalayas to northern Australia, Sri Lanka and Africa. Globally, bananas (*Musa* spp.) form the fourth-most important food crop (Novak *et al.*, 2014). Moreover, the bananas have got much attention for their medicinal, ornamental and socio-economic value (Cordeiro *et al.*, 2004; Aziz *et al.*, 2011; Joe and Sabu, 2016). The two new artificial hybrids of Musaceae, *Musa* × *georgiana* Rich. H. Wallace (Wallace and Hakkinen, 2009) and *Musa* × *formobisiana* H.-L.Chiu, C.-T. Shii & T.-Y.A. Yang. (Chiu *et al.*, 2017) were developed earlier to explore the breeding relevance and ornamental potential value for the banana breeding programmes and landscaping applications.

As part of our studies on reproductive biology of Musaceae with the purpose of breeding evaluation and finding potential ornamental value, intra and intersectional hybridization have been done. The paper focuses on the new artificial hybrid, developed from artificial breeding of the Scarlet Banana *M. coccinea* Andrews (female parent) and *M. haekkinenii* N.S. Lý & Haev. (male parent) which shared the vegetative and floral characters of both the parents and expressed some of its own.

Materials and Methods

The present study was conducted at Calicut University Botanical Garden (11°25'45''N, 75°45'50''E) during 2013–2016. The stigma receptivity of *Musa coccinea* (female parent) and pollen viability of *M. haekkinenii* (male parent) were assessed at different time intervals from anthesis to flower closing using MTT (Dafni *et al.*, 2005) and TTC (Shivanna and Rangaswamy, 1992) tests, respectively. Manual cross pollination was done in a period of maximum stigma receptivity and pollen viability which coincides. The self-compatibility was assessed by a method suggested by Dafni *et al.* (2005). The seed set and seed germination were observed for 3–4 months. The phenological events of new interspecific hybrid were observed with 10 plants of F1 hybrid. Morphometric analysis of vegetative and floral characters was done with a scale and

[1]Corresponding author. Email: msabu9@gmail.com

LEICA M80 stereomicroscope. Colour comparison of new hybrid with that of parents was referred by colour code (Kornerup and Wanscher, 1978). Descriptions of a new hybrid and parent plants were given using INIBAP *Musa* descriptor list (IPGRI-INIBAP/CIRAD, 1996). Photographs were taken with Sony DSC-HX400V digital camera. Voucher specimens were deposited at CALI and MH.

Results and Discussion

Musa × parahaekkinenii K.P. Smisha & M. Sabu, **hybrid nov.**   (**Figs 1&2**).

Diagnosis: *Musa × parahaekkinenii* differs from the female parent *M. coccinea* in having orange red bracts (vs. scarlet red) and presence of yellow fruits (vs. creamy fruits). The *M. × parahaekkinenii* shows distinct characters from the male parent *M. haekkinenii* by the presence of bracts obliquely erect to axis (vs. bracts curving downward). Moreover, the hybrid *M. × parahaekkinenii* exhibits the presence of horn-like appendages on lateral lobes of compound tepal. These appendages are present on the five lobes of compound tepal of *M. coccinea* while totally absent in all lobes of compound tepal of *M. haekkinenii*.

Type: India, Kerala, Malappuram district, Thenhipalam, Calicut University Botanical Garden (11°25'45''N, 75°45'50''E), 09 Dec 2016, *K.P. Smisha* 147908 (*Holotype*: CALI; *Isotype*: MH).

Clump forming; plants slender, herbaceous, suckering freely with 8–10 suckers of 10–15 cm long, oriented vertically. Mature pseudo-stem slender, 90–100 cm high, 13–15 cm in diam. at the base; sap milky. Leaf green, dorsiventral, 100–130 cm length; petiole 28–30 cm long, green with sparse brown blotches at the base, petiole canal margins incurved, narrow, 0.6–0.7 cm wide, scarious, clasping pseudo-stem at the base; lamina oblong-lanceolate, 70–100×23.5–24 cm, apex obtuse, margin corrugated, midrib 70–100×0.6–0.7cm, greyish green adaxially, pale greyish green abaxially, one side rounded and other pointed, adaxial surface dark green and dull, abaxial surface deep green and shiny, insertion point of leaf bases asymmetric on both sides. Inflorescence erect; peduncle 5–8 cm long, 5.0–5.5 cm in diam., glabrous, yellowish cream. Flag leaf with colourful bract like base and leafy apex persistent, 45–55 cm long. Sterile bracts lanceolate, 20–21×4.0–4.5 cm, adaxial surface dull, orange red with yellow tinge at base, abaxial surface shiny, orange red with yellow tinge at base, apex acute, greenish, base greyish orange, small shouldered, persistent. Male bracts lanceolate, 11–12×4.3–4.5 cm, bract lifting one at a time, persistent, adaxial surface dull, orange red, abaxial surface shiny, orange red, apex intermediate, green tinted with yellow, base greyish orange, small shouldered, margin not revolute. Female bracts lanceolate, 12.5–13.5×4.0–4.5 cm, adaxial surface dull, orange red, abaxial surface shiny, orange red, apex acute to obtuse, green tinted with yellow, base greyish orange, small shouldered, margin not revolute, lifting one bract at a time, imbricate. Basal flowers female, yellow, 1 flower per bract, 4–7 cm long. Compound tepal 3–4×1.2–2.3 cm, lower half deep yellow and upper half olive green, ribbed on either side, apex 5-lobed, rounded, with one horn-like appendages on lateral lobes, two lobes larger and exserted, 0.1×0.2 cm, middle and lateral lobes curved backward, 0.1×0.2 cm. Free tepal 3.0–3.8×1.0–1.2 cm, ovate, as long as the style, closely appressed to the stigma, translucent, opaque yellow, margin entire, apex corrugated with short acumen, adaxial surface smooth, abaxial surface ribbed. Staminodes 5, lanceolate, creamy yellow, 0.8–1.7×0.1–0.2 cm. Ovary straight, 2.2–4.0 cm long, lemon yellow, waxy, 3-locular; style straight, 2–3×0.1–0.2 cm, pale yellow with olive green tinge at apex. Stigma yellow, terete, 0.8×0.5 cm. Male flowers yellow, 2 flowers per bract, 4.7–5.0 cm long. Compound tepal 4.5–5 ×1.8–2.3 cm, lower half deep yellow and upper half olive green, ribbed on either side, apex 5-lobed, rounded, with horn-like appendages on lateral lobes, three middle lobes larger and exserted, 0.2×0.1 cm, 3 central lobes curved backward, 0.1×0.1 cm. Free tepal 4.5–4.7×1.0–1.3 cm, ovate, 3/4th of compound tepal, translucent, opaque yellow, margin entire, apex corrugated with short acumen, adaxial surface smooth, abaxial surface ribbed. Stamens 5, 4.2–4.3 cm long; anther greyish yellow, fertile, 1.8–2×0.05–0.1cm; filament

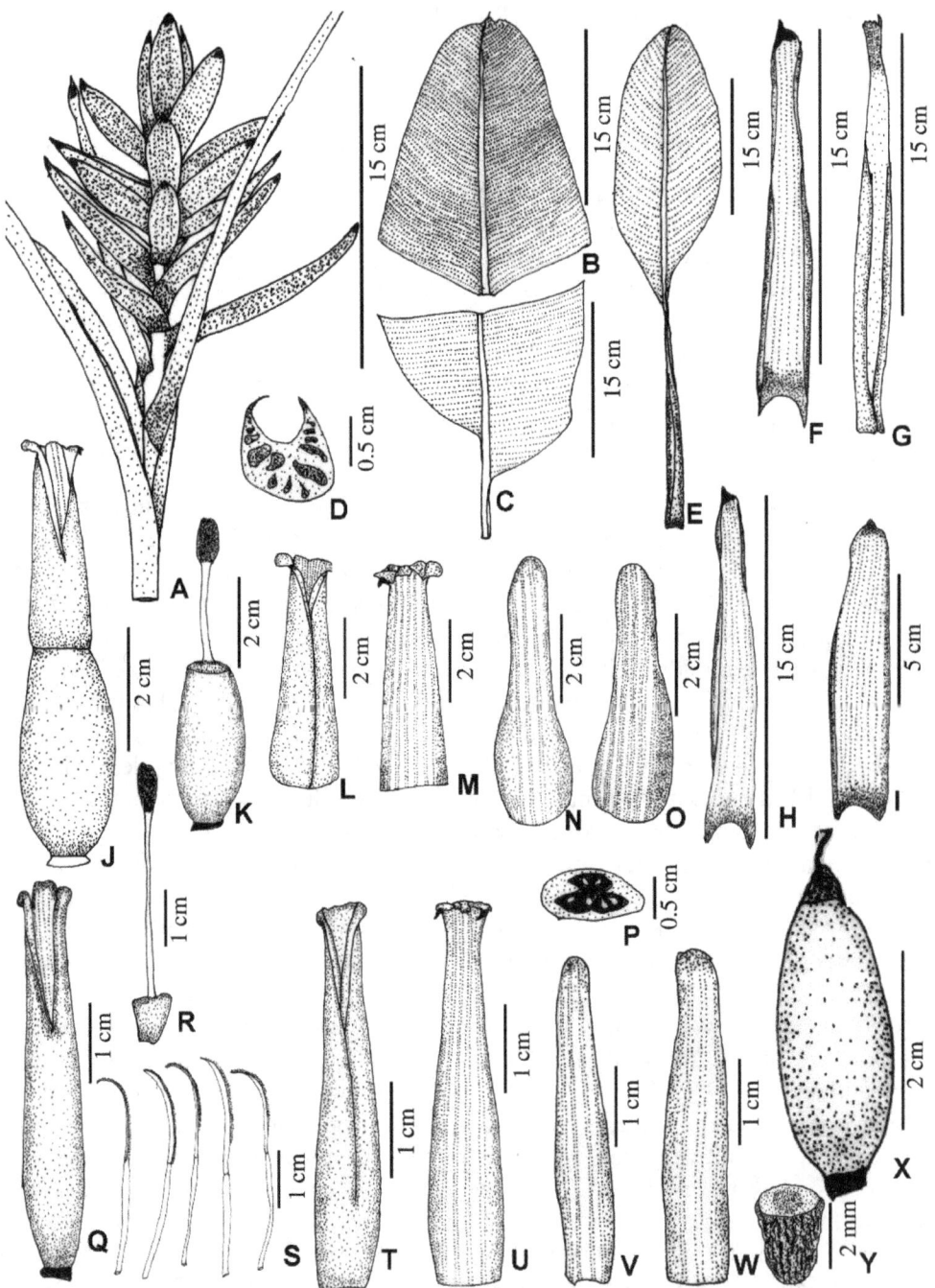

Fig. 1. Illustration of *Musa × parahaekkinenii* **hybrid nov.** A. Inflorescence; B. Leaf apex; C. Leaf base; D. Cross-section of petiole; E. Flag leaf; F&G. Sterile bracts; H. Female bract; I. Male bract; J. Female flower; K. Ovary with style and stigma; L&M. Compound tepals; N&O. Free tepals; P. Cross-section of ovary; Q. Male flower; R. Rudimentary ovary with style and stigma; S. Stamens; T&U. Compound tepals; V&W. Free tepals; X. Fruit; Y. Seed.

Fig. 2. *Musa × parahaekkinenii* **hybrid nov.** A. Habit; B. Inflorescence; C. Corm; D. Pseudostem; E. Cross-section of petiole; F. Leaf apex; G. Leaf base; H. Flag leaf; I. Sterile bracts; J. Female bract; K. Male bract; L. Female flower; M. Ovary with style and stigma; N. Compound tepals; O. Free tepals; P. Staminodes; Q. Cross-section of ovary; R. Male flower; S. Compound tepals; T. Free tepals; U. Stamens; V. Fruit; W. Seed.

1.6–2.5×0.1–0.2 cm, light yellow. Style rudimentary, straight, 3.5–4.4×0.1–0.2 cm, pale yellow with olive green tinge at apex. Stigma rudimentary, terete, yellowish orange, narrowly oblong, 0.6–0.7×0.3–0.4 cm. Ovary rudimentary, oblong, 0.5–0.7×0.4–0.8 cm, pale yellow, waxy. Fruit berry, 1 per bract, mature fruit narrowly oblong, 5.0–6.5 cm long, yellow, not waxy, straight, bear persistent floral relicts; 3.5–4.0 cm long. Seeds numerous, cylindrical, brown, warty, 1.5–2.0 cm long.

Etymology: The epithet has been named for the bract colour that resemblance with the male parent (*M. haekkinenii*) at the first appearance.

Key to the new hybrid *Musa × parahaekkinenii* and its parents

1. Bracts obliquely upward to axis.   2
   – Bracts curving downwards to axis.   *M. haekkinenii*
2. Horn-like appendages only on lateral lobes of compound tepal;   *M. × parahaekkinenii*
   bracts orange red.
   – Horn-like appendages on all lobes of compound tepal; bracts   *M. coccinea*
   scarlet red.

Table 1. Comparison of characters of *Musa × parahaekkinenii* hybrid nov. with its parents.

Characters	*M. haekkinenii*	*M. × parahaekkinenii*	*M. coccinea*
Plant height (cm)	250–300	80–110	150–200
Sap	Watery	Milky	Watery
Sucker number	8–13	9–12	8–10
Petiole canal margins	Wide, erect	Narrow, erect	Narrow, erect
Leaf base	One side rounded, one pointed	Both side pointed	Rounded
Female/Male bract behaviour	Persistent, curving downward to axis	Persistent, obliquely upward to axis	Persistent, obliquely upward to axis
Female/Male bract	Bright orange red on both sides	Bright orange red on both sides	Deep scarlet red on adaxial and abaxial side
Female/Male flowers per bract	1–4	1–2	1–2
Stigma shape	Flat	Narrowly flat	Narrowly flat
Appendage	Absent	Present only on lateral lobes of compound tepal	Present on all lobes of compound tepal
No. of fruit/s per bract and shape	3–4, straight	1, straight	1–2, straight

The petiole canal margins are clasping, ovary deep yellow and the seeds barrel-shaped and warty in the parents and the artificial hybrid.

A new hybrid *Musa × parahaekkinenii*, developed here by manual cross pollination is highly relevant for breeding application and to explore the potential ornamental value. The reciprocal cross between *M. haekkinenii* (female parent) and *M. coccinea* (male parent) also resulted in fruit set and seed production. Hybridization experiments of *M. coccinea* (female parent) with male parents of *M. beccarii*, *M. ornata*, *M. siamensis*, *M. laterita*, *M. velutina* and *M. markkuana* did not result in any fruit set. But the cross between *M. haekkinenii* (female parent) and *M. beccarii* (male parent) resulted in both fruit and seed set.

Musa coccinea is distributed in Indochina and China (Leong-Škorničková and Gallick, 2012). The species is remarkable for its highly attractive scarlet red bracts. The bract is oriented obliquely erect to axis and persistent that improves the potential ornamental value of the species. On the other hand, *Musa haekkinenii* is a native wild banana of Vietnam and known only under cultivation (in Calicut University Botanical Garden, Kerala, India and Suriana Botanic Garden, Penang, Malaysia) today. The centre of origin of the species is northern Vietnam and no wild population of it so far reported. Its existence in the wild is still uncertain. It was recognized as a Data Deficient species according to IUCN Red List categories and criteria (Lý *et al.*, 2012). The plant has a potential ornamental value because of the presence of highly attractive orange red persistent bract.

The new interspecific hybrid *M. × parahaekkinenii* which express intermediate characters of *M. coccinea* (♀) and *M. haekkinenii* (♂) and exhibits new characters also. The basal unisexual female flowers are fertile with receptive stigma and unisexuality is recognized as a unique adaptation for cross pollination. The male flowers are fertile with pollen grain production and recognized as a self-compatible hybrid. However, the new hybrid significantly adds the ornamental value because of the presence of highly attractive persistent bracts and flowers of inflorescence which lasts up to 3–4 months. Currently, many newly explored wild *Musa* species are used as staple food, medicine and ornamentals and also got socio-economic relevance. So, the interspecific hybridization between wild species has an immense value for breeding purposes and to explore the aspects of genetic variability which form the basis of genetic diversity.

Acknowledgements

The authors thank Kerala State Council for Science, Technology and Environment (KSCSTE), providing necessary funds for research work (Order No. 402/2015/KSCSTE dated 18.08.2015).

References

Aziz, N.A.A., Ho, L.-H., Azanari, A., Bhat, R., Cheng, L.-H. and Ibrahim, M.N.M. 2011. Chemical and functional properties of the native banana (*Musa acuminata × balbisiana* Colla cv. Awak) pseudo-stem and pseudo-stem tender core flours. Food Chem. **128**: 748–753.

Chiu, H.-L., Shii, C.-T. and Yang, T.Y.A. 2017. *Musa × formobisiana* (Musaceae), a new interspecific hybrid banana. Taiwania **62**: 147–150.

Cordeiro, N., Belgacem, M.N., Torres, L.C. and Moura, J.C.V.P. 2004. Chemical composition and pulping of banana pseudo-stems. Ind. Crops Prod. **19**: 147–154.

Dafni, A., Kevan, P.G. and Husband, B.C. 2005. Practical Pollination Biology. Enviroquest Ltd., Canada, pp. 1–590.

IPGRI-INIBAP/CIRAD. 1996. Description for Bananas (*Musa* spp.). International Plant Genetic Resources Institute, Rome, Italy/ International Network for the Improvement of Banana and Plantain, Montpellier, France/ Centre de Cooperation Internationale en Rechereche Agronomique pour le Development, Montpellier, France, pp. 1–58.

Joe, A. and Sabu, M. 2016. Wild ornamental bananas in India: an overview. Sourth Ind. J. Biol. Sci. **2**: 213–221.

Kornerup, A. and Wanscher, J.H. 1978. Methuen Handbook of Colour, 3rd edn. Methuen, London, pp. 1–252.

Leong-Škorničková, J. and Gallick, D. 2012. The Ginger Garden. National Parks Board Singapore Botanic Garden, Singapore, 114 pp.

Lý, N.S., Lê, C.-K., Triệu, T.-D., Haevermans, A., Lowry II, P.P. and Haevermans, T. 2012. A distinctive new species of wild banana (*Musa*, Musaceae) from northern Vietnam. Phytotaxa **75**: 33–42.

Novák, P., Hřibová, E., Neumann, P., Koblížková, A., Doležel, J. and Macas, J. 2014. Genome wide analysis of repeat diversity across the family Musaceae. PLoS ONE **9**: e98918.

Shivanna, K.R. and Rangaswamy, N.S. 1992. Pollen Biology: A Laboratory Manual. Narosa Publishing House, New Delhi, pp.1–199.

Wallace, R. and Häkkinen, M. 2009. *Musa* × *georgiana,* a new intersectional hybrid banana with edible banana breeding relevance and ornamental potential. Nordic J. Bot. **27**: 182–185.

20

INCLUSION OF *KICKXIA ABHAICA* D.A. SUTTON IN THE GENUS *NANORRHINUM* (PLANTAGINACEAE): EVIDENCE FROM ITS NUCLEAR RIBOSOMAL DNA SEQUENCES

M. Ajmal Ali[1]

Department of Botany and Microbiology, College of Science, King Saud University, Riyadh-11451, Saudi Arabia

Keywords: *Kickxia abhaica*; Antirrhineae; *Nanorrhinum abhaicum*; ITS; nrDNA; Saudi Arabia.

Abstract

The nuclear ribosomal DNA (nrDNA) internal transcribed spacers (ITS) sequences is extensively used in the plant molecular phylogenetics for plant taxonomic identification and DNA barcoding purposes because the nrDNA ITS gene is easy to amplify by using the universal primers, its length is shorter and thus easy to sequence, and has strong discrimination power to distinguish the taxon at the species level. The present molecular phylogenetic analysis of ITS nrDNA sequences focuses to determine the taxonomic status of an unresolved endemic taxon *Kickxia abhaica* D.A. Sutton (Family Plantaginaceae, tribe Antirrhineae) reported from Saudi Arabia. The analysis supports the transfer of *K. abhaica* under the genus *Nanorrhinum*.

Introduction

The tribe Antirrhineae which comprises ca. 30 genera (Sutton, 1988) has undergone several taxonomic changes during last two decades. The genus *Kickxia* Dumort. (Family Plantaginaceae, tribe Antirrhineae) comprises ca. 25 accepted species (APG III, 2009). Based on the mode of dehiscence of capsule, the genus *Kickxia* has been divided into sections i.e. *Kickxia* sect. *Kickxia* and *Kickxia* sect. *Valvatae* (Sutton, 1988). The sections *Kickxia* sect. *Kickxia* and *Kickxia* sect. *Valvatae* were raised to the rank of subgenera (Smith, 1973). The species with valvate capsules were treated under *Pogonorrhinum* and *Nanorrhinum* (Betsche, 1984). Ghebrehiwet (2001) considered *Kickxia* and *Nanorrhinum* as two distinct genera on the basis of morphological analysis. The molecular phylogeny of mediterranean genera *Chaenorhinum*, *Kickxia* and *Nanorrhinum* based on nrDNA ITS and *rpl32-trn*L sequence data also supports the recognition of the clade comprising *Kickxia* sect. *Valvatae* as *Nanorrhinum*; as a result, new combinations i.e. *Nanorrhinum petranum* (Danin) Yousefi & Zarre, *Nanorrhinum judaicum* (Danin) Yousefi & Zarre and *Nanorrhinum scariosepalum* (Tackh. & Boulos) Yousefi & Zarre were established from *Kickxia petrana* Danin, *Kickxia judaica* Danin and *Kickxia scariosepala* Tackh. & Boulos, respectively (Yousefi *et al.*, 2016).

The genus *Kickxia* in Saudi Arabia is represented by nine species and one subspecies [*Kickxia abhaica* D.A. Sutton, *K. acerbiana* (Boiss.) Tackh. & Boulos, *K. aegyptiaca* (L.) Nab., *K. collenetteana* D.A. Sutton, *K. corallicola* D.A. Sutton, *K. elatine* subsp. *crinita* Greuter, *K. hastata* (R.Br. *ex* Benth.) Dandy, *K. petiolata* D.A. Sutton, *K. pseudoscoparia* V.W. Smith and *K. scalarum* D.A. Sutton] described under the family Scrophulariaceae (Chaudhary, 2001), out of which the taxonomic status of *K. abhaica* D.A. Sutton, *K. acerbiana* (Boiss.) Tackh. & Boulos and *K. hastata* (R.Br. *ex* Benth.) Dandy is still unresolved (http://www.theplantlist.org/), *K. abhaica* D.A. Sutton [Rev. Antirrhinea: 241 (1988). Plate Scroph. 17.] have been reported as

[1]E-mail: ajmalpdrc@gmail.com

endemic to Saudi Arabia (Chaudhary, 2001). The present study aims to resolve the taxonomic status of the *K. abhaica* based on the molecular phylogenetic analysis of ITS nrDNA sequences.

Materials and Methods

Collection of the leaf material of Kickxia abhaica:

The leaf material of *K. abhaica* was collected from the specimen [Dharb-Abha Road, 5-4-1982, S. Chaudhary 3907 (KSUH)] deposited at the Herbarium (Department of Botany and Microbiology, College of Science, King Saud University, Riyadh, Saudi Arabia). A total of 16 species of *Kickxia* was employed in this study (Table 1). The taxonomic identification of the herbarium specimens were reconfirmed with the taxonomic description mentioned in recent Flora of the Kingdom of Saudi Arabia (Chaudhary, 2001).

Table 1. The GenBank accessions of the ingroup and outgroup taxon included in the molecular phylogenetic analysis of *Kickxia abhaica*.

No.	Taxon	GenBank Acc. No.
Ingroup		
1.	*Nanorrhinum cabulicum* (Benth.) Podlech & Iranshahr	KT031916
2.	*Kickxia sagittata* (Poir.) Rothm.	KT031902
3.	*K. scoparia* (Brouss. ex Spreng.) G.Kunkel & Sunding	KT031903
4.	*K. urbanii* (Pit.) K.Larsen	KT031915
5.	*K. scariosepala* Täckh. & Boulos	KT031911
6.	*K. macilenta* (Decne.) Danin	KT031908
7.	*K. petrana* Danin	KT031909
8.	*K. judaica* Danin	KT031907
9.	*Kickxia lanigera* (Desf.) Hand.-Mazz.	KX061033
10.	*K. spuria* (L.) Dumort.	KT031914
11.	*K. sieberi* (Rchb.) Dörfl. & Allan	KT031912
12.	*K. cirrhosa* (L.) Fritsch	KT031896
13.	*K. aegyptiaca* (L.) Nab.	KT031905
14.	*K. elatine* (L.) Dumort.	KT031898
15.	*K. commutate* (Bernh. ex Rchb.) Fritsch	KT031897
16.	*K. abhaica* D.A. Sutton	MH628533
	[= ***Nanorrhinum abhaicum* (D.A. Sutton) Ajmal Ali *comb. nov.*]**	
Outgroup		
17.	*Anarrhinum bellidifolium* (L.) Willd.	AY878116

Extraction of genomic DNA, amplification and sequencing of nrDNA ITS gene:

The leaf material was crushed with liquid nitrogen using 'Qiagen Tissue Lyser' (# 85300). The robotic workstation 'QIAcube' (# 9001292) using 'DNeasy Plant Mini Kit' (# 69104) was used for automated purification of the total genomic DNA. The nuclear ribosomal DNA ITS sequences (ITS1-5.8S and ITS2) were amplified in the thermal cycler (Applied Biosystems Veriti) via polymerase chain reaction using the primers (White *et al.*, 1990) [forward primer ITS1 (5' GTCCACTGAACCTTATCATTTAG3') and the reverse primer ITS4 (5'TCCTCCGCTTATT GATATGC3')] and PCR Mix (# K-2011, Bioneer, Daejeon, Republic of Korea). The DNA sequencing of the amplified product was performed using kit (# 4337455, BigDye Terminator cycle sequencing kit, Perkin-Elmer, Applied Biosystems) in DNA Analyzer (Perkin- Elmer, Applied Biosystems, # ABI PRISM 3730XL).

Molecular phylogenetic analysis of the nrDNA ITS gene sequences:

The nrDNA ITS sequences of a total number of 16 species of *Kickxia s.s.* and *s.l.* and outgroup sequence (Table 1) were retrieved from NCBI GenBank. The ITS sequences of nrDNA of *Anarrhinum bellidifolium* (GenBank accession No. AY878116) was used as outgroup in the molecular phylogenetic analysis because the genus *Anarrhinum* shows close relationships to the genus *Kickxia* (Yousefi *et al.*, 2016). The alignment software 'CLUSTAL X v.1.81' (Thompson *et al.*, 1997) was used to align the FASTA format DNA sequences. The parsimony (maximum parsimony, MP) (Nei and Kumar, 2000; Eck and Dayhoff, 1996) analysis using bootstrap method (Felsenstein, 1985) and maximum likelihood (ML) analysis using maximum composite likelihood method (Tamura *et al.*, 2004) were used to conduct the molecular phylogenetic analyses using the molecular phylogenetic analysis software MEGA X (Kumar *et al.*, 2018).

Results and Discussion

The aligned nrDNA ITS data (ITS1, 5.8S, and ITS2 region) matrix was 622 bp (base pair) long. The most parsimonious tree out of nine parsimonious trees (length = 84) showed consistency index (CI) 0.781 and retention index (RI) 0.932. The ITS region (ITS1-5.8S-ITS2) of *K. abhaica* possessed 613 bp [ITS1: 228 bp, GC content 69%; 5.8S: 164 bp, GC content 54%; ITS2: 221 bp, GC content 71%].

The present molecular phylogenetic analysis of nrDNA ITS sequences revealed that *Kickxia s.l.* is monophyletic and sister to *Kickxia s.s.* The maximum parsimony phylogenetic tree (Fig. 1) showed two main clades i.e. *Kickxia s.s.* clade (BS 96%) and *Nonorrhinum* clade (BS 100%). *K. abhaica* nested within the *Kickxia s.l./Nonorrhinum* clade (BS 90%). The *Kickxia s.l.* (*K. scoparia - K. urbani - K. sagittata*) clade forms a distinct group (BS 90%). The ML tree with the highest log likelihood (-1294.69) recovered phylogenetic tree topology similar to MPT (Fig 1).

The tribe Antirrhineae (under Scrophulariaceae *s.l.*), with c. 300 species distributed in c. 30 genera constitutes a major clades of Plantaginaceae (Albach *et al.*, 2005). The member of the tribe Antirrhineae are characterized by their herbaceous habit; two-lipped tubular corolla, 3-lobed lower lip and 2-lobed upper lip, gibbose, sometimes spurred at the base; 5 epipetalous stamens out of which 2 or 4 fertile, 2-carpelled fruits, operculate / valvate capsules (Sutton, 1988), and unique antirrhinosides / iridoid glycosides (Beninger *et al.*, 2008). The systematic position of both the tribe and genera of the tribe Antirrhineae has been much debated (Ghebrehiwet *et al.*, 2000), and the generic limits is still unresolved especially in the case of the genera *Chaenorhinum*, *Kickxia* and *Nanorrhinum* (Ghebrehiwet *et al.*, 2000; Albach *et al.*, 2005).

The morphological characteristics of taxon at lower level vary under different geographical and environmental condition; hence, requires sufficient taxonomic expertise for taxon identification based on morphology. In contrast, the DNA sequences have least or hardly influence by the geographical or environmental condition, and even remain unchanged during the developmental stages; therefore, the DNA barcode sequence such as ITS, ycf5, *rbcL, mat*K, *rpo*C1, *psb*A-*trn*H, *ndh*F, *trn*L-F, and *rps*16 based species identification together with morphological features gaining wide acceptance recently (Marcon *et al.*, 2005; Liu *et al.*, 2011; Rai *et al.*, 2012; Ali *et al.*, 2014). The ML tree showed two main clades i.e. *Kickxia s.s.* clade (BS 99%) and *Nonorrhinum* clade (BS 100%). *K. abhaica* nested within the *Kickxia s.l./Nonorrhinum* Clade (BS 71%), the *Kickxia s.l.* (*K. sagittata- K. scoparia-K. urbani*) clade forms a distinct group (BS 90%). Previously, *K. scoparia, K. urbani* and *K. sagittata* were recognized as *Nanorrhinum*

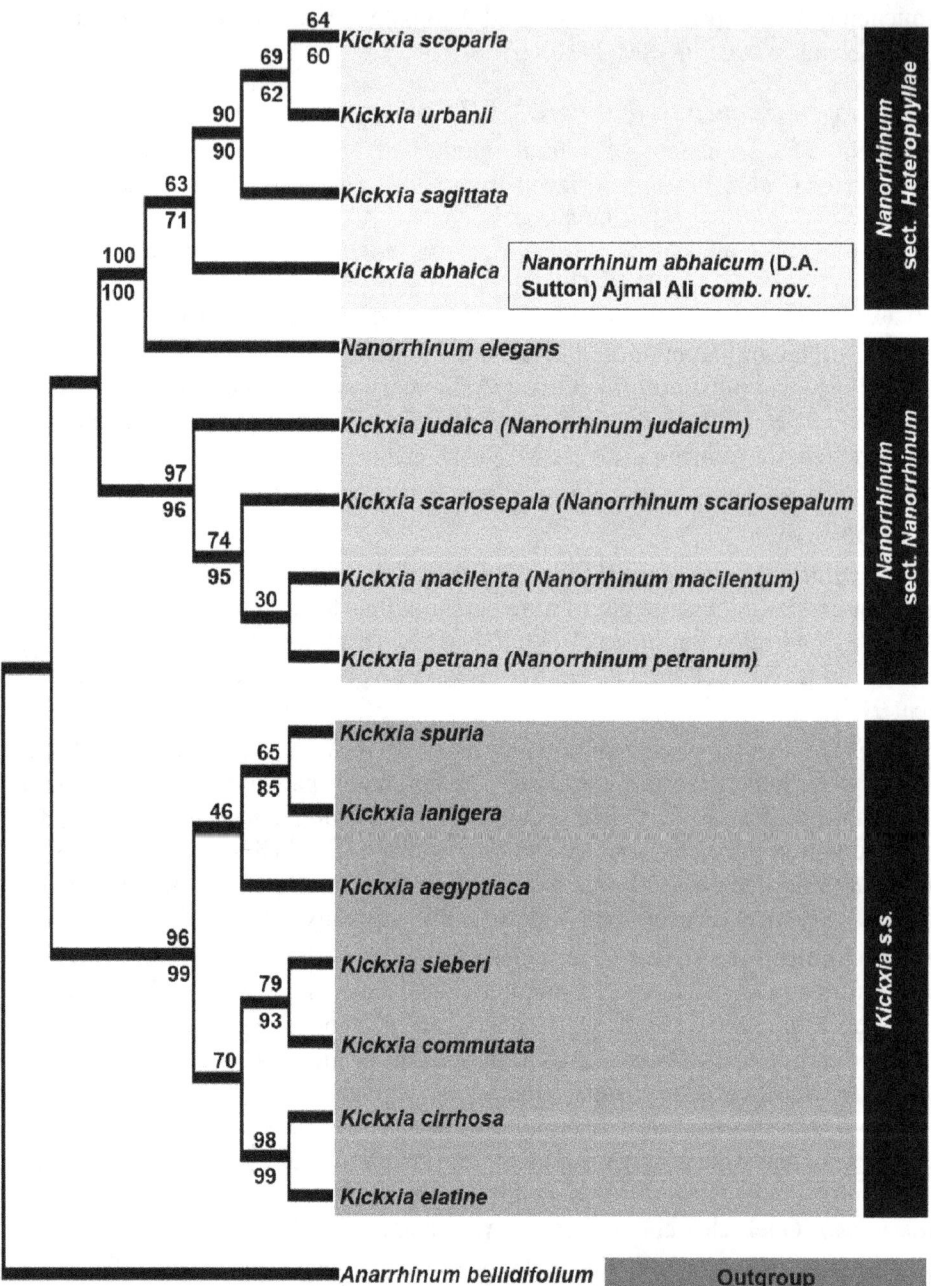

Fig. 1. The phylogenetic tree showing the systematic position of *Kickxia abhaica* [= *Nanorrhinum abhaicum* (D.A. Sutton) Ajmal Ali *comb. nov.*]. The phylogenetic analysis (1000 bootstrap replicates) was inferred using the Maximum Parsimony method. The numbers at the nodes are the bootstrap supports in MP (above) and ML (below) analysis.

(Smith, 1973) under the sect. *Heterophyllae* (Yousefi *et al.*, 2016) or as *Pogonorrhinum* (Betsche, 1984). The taxonomic status of *K. colletteana* (branches prostrate spreading, rigid, leafy; leaves all elliptic to oblong), *K. corallicola* (branches flexuous, tangled; petiole long, capillary, twining), *K. hastata* (annual delicate herb*)*, *K. petiolata* (leaves homomorphic, without any basal lobe; petioles becoming thickened woody; spur coming out from corolla base) and *K. scalarum* (petioles prominent capillary, often twining; corolla drying dark) reported from Saudi Arabia are unresolved, and its DNA sequence for any gene are not available in the GenBank. Therefore, the DNA sequencing of these taxon are required to know its taxonomic status within the tribe Antirrhineae. Moreover, the molecular phylogenetic analysis of nrDNA ITS sequence of *K. abhaica* [which was described as endemic to Saudi Arabia (Chaudhary, 2000)] supports its transfer to the genus *Nanorrhinum,* and thus the proposed new combinations in *Nanorrhinum* (new generic record for Saudi Arabia) is as follows.

New combination in *Nanorrhinum*

Nanorrhinum abhaicum (D.A. Sutton) Ajmal Ali, **comb. nov.**

Basionym: *Kickxia abhaica* D.A. Sutton [Rev. Antirrhinea: 241 (1988). Plate Scroph. 17.]

Acknowledgement

Research supported by the King Saud University, Deanship of Scientific Research, College of Science, Research Center.

References

Albach, D.C., Meudt, H.M. and Oxelman, B. 2005. Piecing together the 'new' Plantaginaceae. Am. J. Bot. **92**: 297–315.

Ali, M.A., Gábor, G., Norbert, H., Balázs, K., Al-Hemaid, F.M.A., Pandey, A.K. and Lee, J. 2014. The changing epitome of species identification - DNA barcoding. Saudi J. Biol. Sci. **21**(3): 204–231.

APG III (Angiosperm Phylogeny Group). 2009. An update of the Angiosperm Phylogeny Group classification for the orders and families of flowering plants: APG III. Bot. J. Linn. Soc. **161**: 105–121.

Beninger, C.W., Cloutier, R.R. and Grodzinski, B. 2008. The iridoid glucoside, antirrhinoside from *Antirrhinum majus* L. has differential effects on two generalist insect herbivores. J. Chem. Ecol. **34**: 591–600.

Betsche, L. 1984. Taxonomische Untersuchungen an *Kickxia* Dumortier (*s.1.*). Die neuen Gattungen *Pogonorrhinum* n. gen. und *Nanorrhinum* n. gen. Cour. Forsch. Inst. **71**: 125–142.

Chaudhary, S. (Ed.) 2001. Flora of the Kingdom of Saudi Arabia. Ministry of Agriculture and Water, National Herbarium, National Agriculture and Water Research Center, Riyadh, Saudi Arabia, Vol. **II**(2), pp. 435–439.

Eck, R.V. and Dayhoff, M.O. 1966. Atlas of Protein Sequence and Structure. National Biomedical Research Foundation, Silver Springs, Maryland, USA.

Felsenstein, J. 1985. Confidence limits on phylogenies: An approach using the bootstrap. Evolution **39**: 783–791.

Ghebrehiwet, M. 2001. Taxonomy, phylogeny and biogeography of *Kickxia* and *Nanorrhinum* (Scrophulariaceae). Nordic J. Bot. **20**: 655–690.

Kumar, S., Stecher, G., Li, M., Knyaz, C. and Tamura, K. 2018. MEGA X: Molecular Evolutionary Genetics Analysis across computing platforms. Mol. Biol. Evol. **35**: 1547–1549.

Liu, C., Liang, D., Gao, T., Pang, X., Song, J., Yao, H., Han, J., Liu, Z., Guan, X., Jiang, K., Li, H. and Chen, S. 2011. PTIGS-IdIt, a system for species identification by DNA sequences of the *psb*A-*trn*H intergenic spacer region. BMC Bioinformatics **12**: S4.

Marcon, A.B., Barros, I.C. and Guerra, M. 2005. Variation in chromosome numbers, CMA bands and 45S rDNA sites in species of *Selaginella* (Pteridophyta). Ann. Bot. **95**: 271–276.

Nei, M. and Kumar, S. 2000. Molecular Evolution and Phylogenetics. Oxford University Press, New York.

Rai, P.S., Bellampalli, R., Dobriyal, R.M., Agarwal, A., Satyamoorthy, K. and Narayana, D.A. 2012. DNA barcoding of authentic and substitute samples of herb of the family Asparagaceae and Asclepiadaceae based on the ITS2 region. J. Ayurveda Integr. Med. **3**: 136–140.

Smith, V.A. 1973. A revision of the genus *Kickxia* with particular reference to the section *Heterophyllae* in the Canary Islands. M. Sc. thesis, University of Reading, UK.

Sutton, D.A. 1988. A revision of the tribe Antirrhineae. Oxford Univ. Press.

Tamura, K., Nei, M. and Kumar, S. 2004. Prospects for inferring very large phylogenies by using the neighbor-joining method. Proc. Nat. Acad. Sci. (USA) **101**: 11030–11035.

Thompson, J.D., Gibson, T.J., Plewniak, F., Jeanmougin, F. and Higgins, D.G. 1997. The CLUSTAL_X windows interface: Flexible strategies for multiple sequence alignment aided by quality analysis tools. Nucleic Acids Res. **24**: 4876–4882.

Yousefi, N., Zarre, S. and Heubl, G. 2016. Molecular phylogeny of the mainly Mediterranean genera *Chaenorhinum*, *Kickxia* and *Nanorrhinum* (Plantaginaceae, tribe Antirrhineae), with focus on taxa in the Flora Iranica region. Nordic J. Bot. **34**: 455–463.

PERMISSIONS

LIST OF CONTRIBUTORS

Yourang Hwang and Man Kyu Huh
Department of Molecular Biology, Dong-eui University, 995 Eomgwangno, Busanjin-gu, Busan 614-714, Korea

Aleya Ferdousi, Md. Oliur Rahman and Md. Abul Hassan
Department of Botany, University of Dhaka, Dhaka-1000, Bangladesh

Leila Samiei, Mahnaz Kiani and Homa Zarghami
Department of Ornamental Plants, Research Center for Plant Sciences, Ferdowsi University of Mashhad, Mashhad, Iran

Farshid Memariani and Mohammad Reza Joharchi
Department of Botany, Research Center for Plant Sciences, Ferdowsi University of Mashhad, Mashhad, Iran

M. Ajmal Ali
Department of Botany and Microbiology, College of Science, King Saud University, Riyadh-11451, Saudi Arabia

Joongku Lee
Department of Environment and Forest Resources, Chungnam National University, 99 Daehak-ro, Yuseonggu, Daejeon 34134, South Korea

M. Oliur Rahman
Department of Botany, University of Dhaka, Dhaka 1000, Bangladesh

Fahad S. M. Al-Anazi, Fahad M. A., Al-Hemaid, A. A. Hatamleh and Changyoung Lee
International Biological Material Research Center, Korea Research Institute of Bioscience and Biotechnology, 111 Gwahangno, Yuseong-gu, Daejeon 305 806, South Korea

B. J. Mylliemngap and A. Bhattacharjee
Department of Biotechnology and Bioinformatics, North Eastern Hill University, Shillong 793002, Meghalaya, India

M. Ajmal Ali and Fahad M. Al-Hemaid
Department of Botany and Microbiology, College of Science, King Saud University, Riyadh 11451, Saudi Arabia

Ritesh K. Choudhary, Joongku Lee and Soo-Yong Kim
International Biological Material Research Center, Korea Research Institute of Bioscience and Biotechnology, Daejeon-305 806, South Korea

M. A. Rub
National Herbarium & Genebank, National Agriculture & Animal Resources Research Center, Riyadh-11484, Saudi Arabia

Shamim Shamsi, Sarowar Hosen, Md. Al-Mamun and Momtaz Begum
Department of Botany, University of Dhaka, Dhaka-1000, Bangladesh

Fahad M. A. Al-Hemaid and M. Ajmal Ali
Department of Botany and Microbiology, College of Science, King Saud University, Riyadh 11451, Saudi Arabia

Joongku Lee
Department of Environment and Forest Resources, Chungnam National University, 99 Daehak-ro, Yuseong-gu, Daejeon 34134, South Korea

Soo-Yong Kim
International Biological Material Research Center, Korea Research Institute of Bioscience and Biotechnology, Daejeon 305 806, South Korea

M. Oliur Rahman
Department of Botany, University of Dhaka, Dhaka 1000, Bangladesh

Reza Sheikhakbari-Mehr
Department of Biology, Faculty of Science, University of Qom, Qom, Iran

Ali Asghar Maassoumi
Botany Division, Research Institute of Forests and Rangelands, Tehran, Iran

Shahrokh Kazempour Osaloo
Department of Botany, Faculty of Biological Sciences, Tarbiat Modares University, Tehran, Iran

Mohammad Zashim Uddin and Md. Abul Hassan
Department of Botany, University of Dhaka, Dhaka 1000, Bangladesh

Yong Yang
State Key Laboratory of Systematic and Evolutionary Botany, Institute of Botany, Chinese
Academy of Sciences, 20 Nanxincun, Xiangshan, Beijing 100093, China

M. Oliur Rahman and Md. Zahidur Rahman
Department of Botany, University of Dhaka, Dhaka-1000, Bangladesh

Sonia Khan Sony and Mohammad Nurul Islam
Department of Botany, University of Barisal, Barisal 8200, Bangladesh

M. Ajmal Ali
Department of Botany and Microbiology, College of Science, King Saud University, Riyadh 11451, Kingdom of Saudi Arabia

Joongku Lee and Soo-Yong Kim
International Biological Material Research Center, Korea Research Institute of Bioscience and Biotechnology, 125 Gwahak-ro, Yuseong-gu, Daejeon 305-806, South Korea

Sang-Hong Park and Fahad M. A. Al-Hemaid
International Biological Material Research Center, Korea Research Institute of Bioscience and Biotechnology, 125 Gwahak-ro, Yuseong-gu, Daejeon 305-806, South Korea
Division of Plant Management, National Institute of Ecology, Choongnam, Secheon-gun, Maseo-myeon, Geumgang-ro, 1210, 325-813, South Korea

Sherif M. Sharawy
Botany Department, Faculty of Science, Ain Shams University, Cairo, Egypt
Biology department, Faculty of Science, Hail University, Hail, Saudi Arabia

Abdelfattah Badr
Botany and Microbiology Department, Faculty of Science, Helwan University, Cairo, Egypt

D. P. G. Shashika K. Guruge
Department of Botany, Faculty of Science, University of Peradeniya, Sri Lanka

Deepthi Yakandawala
Postgraduate Institute of Science, University of Peradeniya, Sri Lanka

Kapila Yakandawala
Department of Horticulture & Landscape Gardening, Faculty of Agriculture & Plantation Management, Wayamba University of Sri Lanka

Fahad M. A. Al-Hemaid and M. Ajmal Ali
Department of Botany and Microbiology, College of Science, King Saud University, Riyadh 11451, Saudi Arabia

Joongku Lee
International Biological Material Research Center, Korea Research Institute of Bioscience and Biotechnology, Daejeon 305 806, South Korea

Gábor Gyulai
Institute of Genetics and Biotechnology, St. István University, Gödöllo H-2103, Hungary

Arun K. Pandey
Department of Botany, University of Delhi, Delhi 110007, India

Dhafer Ahmed Alzahrani
Department of Biological Sciences, Faculty of Science, King Abdulaziz University, Jeddah, Saudi Arabia

Enas Jameel Albokhari
Department of Biological Sciences, Faculty of Applied Sciences, Umm Al-Qura University, Makkah, Saudi Arabia

Uzzal Hossain and M. Oliur Rahman
Department of Botany, University of Barisal, Barisal-8200, Bangladesh
Department of Botany, University of Dhaka, Dhaka-1000, Bangladesh

S. N. P. Suriyanti and G. Usup
School of Environmental Science and Natural Resources, Faculty Science and Technology, Universiti Kebangsaan Malaysia, 43600 Bangi, Selangor, Malaysia
UTM Ocean Thermal Energy Centre (OTEC), Ground Floor, Block Q, Universiti Teknologi Malaysia, Jalan Sultan Yahya Petra, 54100 Kuala Lumpur, Malaysia

Shawkat Mahmoud Ahmed
Biology Department, Faculty of Education, Ain Shams University, Cairo, Egypt

M. Ajmal Ali
Department of Botany and Microbiology, College of Science, King Saud University, Riyadh-11451, Saudi Arabia

Index

www.ingramcontent.com/pod-product-compliance
Lightning Source LLC
Chambersburg PA
CBHW080401190526
45161CB00003B/101